Be supported by half a million makers

Put the power of community behind your ideas and bring them to life faster. Avnet connects you to engineers, innovators and tech partners to speed the development and delivery of your innovative, world-changing products.

Jumpstart your next project at
avnet.me/communities

CONTENTS

Make: **Volume 57** June / July 2017

SPECIAL SECTION
Board Guide 2017

ON THE COVER:
Adafruit's Limor Fried and the new Circuit Playground.

Photo: Hep Svadja.
Illustration: Rob Nance

Put a Board on It 24
The current crop of microcontrollers vies for your attention.

Extraordinary Boards 28
These eight new devices push the limits of what you can do with DIY tech.

Open for Business 32
Running an open source hardware and software company has its rewards. But it also has its challenges.

The Ripe Stuff 36
With open source ideals and an engineering genius at the helm, Adafruit is a maker revolution in manufacturing — and so much more.

An Open Chip 42
The journey to develop an open source instruction set for maker-made processors.

Every Child an Inventor 45
The Micro:bit Foundation is spreading its educational mission around the world.

Review: Voltera V-One 46
Make custom PCBs at home with less mess.

Special Pullout
Make: Handy-Dandy Board Guide 2017
With classics and newcomers alike, we've compiled the important specs you need to find the right board for you.

We Are All Makers

makezine.com/57

10

66

72

COLUMNS

Reader Input 06
Thoughts, tips, and musings from readers like you.

Welcome: Overboard 08
So many crowdfunded boards to explore, so little time.

Made on Earth 10
Backyard builds from around the globe.

FEATURES

Clearing the Air 18
The FAA's drone regulations could hamper your flying endeavors. Here's how.

Game Changers 20
alt.ctrl.GDC showcases wild and weird video game control schemes that create a whole new kind of immersion.

Maker ProFile: Ran Ma 22
The biomedical engineer is developing smart socks to help monitor diabetic health.

SKILL BUILDER

Digital Stencil Design 48
Use a laser cutter or CNC machine to transform files into spray-paintable templates.

One Small Step for Robots 52
Program a quadruped with Arduino to get your bot moving.

PROJECTS

Camera-Mounted "Pinoculars" 54
Snap and view zoomed-in photos through your binoculars, with a Raspberry Pi and Pi Camera.

Pedal to the Metal 60
Build this open source stomp box, and rock out with the best of them.

Cat Activated Laser 62
This hyperintelligent toy uses furry face recognition to entertain your pet.

Tricolor Twinkle Lights 66
Brighten your July 4th festivities with jars of sequentially fading strands of LEDs.

Remaking History: The Birth of Medical Hygiene 70
Celebrate Ignaz Semmelweis and his maternity ward revelation by whipping up some bars of homemade soap.

Traveler's Notebook 72
Craft your own cover and paper inserts for a customizable Midori-style journal.

Air Hockey Robot 74
Build your own open source, 3D-printed opponent.

Satisfying Spinners 77
From duct tape to bronze, handmade bearing-based fidget toys are all the rage.

Poseable Papercraft 78
Print, fold, and glue your very own Makey robot mascot.

Light-Up Copter Launcher 80
An LED illuminates this simple cardboard and craft stick contraption.

1+2+3: Portable Rally Sign 82
Make a tall and tote-worthy double-sided sign from reusable and reused materials.

TOOLBOX

3D Printer Review: MakerBot Replicator+ 84
Continuous quality prints indicate the line has fixed its breakdown woes.

SHOW & TELL

Show & Tell 88
Dazzling projects from inventive makers like you.

52

62

60

> "I want to show people that engineering isn't something cold and calculated. Thinking like an engineer is a beautiful and fascinating way to see the world, too."
> — Limor "Ladyada" Fried

EXECUTIVE CHAIRMAN & CEO
Dale Dougherty
dale@makermedia.com

CFO & PUBLISHER
Todd Sotkiewicz
todd@makermedia.com

VICE PRESIDENT
Sherry Huss
sherry@makermedia.com

EDITORIAL

EXECUTIVE EDITOR
Mike Senese
mike@makermedia.com

PROJECTS EDITOR
Keith Hammond
khammond@makermedia.com

SENIOR EDITOR
Caleb Kraft
caleb@makermedia.com

MANAGING EDITOR, DIGITAL
Sophia Smith

PRODUCTION MANAGER
Craig Couden

COPY EDITOR
Laurie Barton

EDITORIAL INTERN
Jordan Ramee

CONTRIBUTING EDITORS
William Gurstelle
Charles Platt
Matt Stultz

DESIGN, PHOTOGRAPHY & VIDEO

ART DIRECTOR
Juliann Brown

PHOTO EDITOR
Hep Svadja

SENIOR VIDEO PRODUCER
Tyler Winegarner

LAB/PHOTO INTERN
Sydney Palmer

MAKEZINE.COM

WEB/PRODUCT DEVELOPMENT
David Beauchamp
Rich Haynie
Bill Olson
Kate Rowe
Sarah Struck
Clair Whitmer
Alicia Williams

CONTRIBUTING WRITERS
Alasdair Allan, Philip J. Angileri, Krste Asanovic, Gareth Branwyn, Sam Brown, Tim Deagan, DC Denison, Kelly Egan, ElectroSmash, Josh Elijah, Jose Julio and Juan Pedro of JJRobots, Linn from Darbin Orvar, Lisa Martin, Forrest M. Mims III, Zach Shelby, Paul Spinrad, Marc De Vinck, John M. Wargo, Gordon Williams, Josh Williams

CONTRIBUTING ARTISTS
Monique Convertito, Vlad Cristea and Graphic Burger for Brusher font, Mike Gray, Rob Ives, Rob Nance, Peter Strain

ONLINE CONTRIBUTORS
Duane Benson, Biohacking Safari, Becky Cater, Kathy Ceceri, Jon Christian, Jeremy Cook, Oliver Damian, Natasha Dzurny, Sean Fairburn, Giedrius Kavaliauskas, Ted Kinsman, Art Krumsee, Jeanne Loveland, Matt McEntee, Goli Mohammadi, Barett Poley, Andrew Salomone, Zachary Samalonis, Craig Schwartz, Andrew Terranova, Sarah Vitak, Michael Weinberg, Glen Whitney, Jennifer Zimmerman

PARTNERSHIPS & ADVERTISING
makermedia.com/contact-sales or sales@makezine.com

DIRECTOR OF PARTNERSHIPS & PROGRAMS
Katie D. Kunde

STRATEGIC PARTNERSHIPS
Cecily Benzon
Allison Davis
Brigitte Mullin

DIRECTOR OF MEDIA OPERATIONS
Mara Lincoln

BOOKS

PUBLISHER
Roger Stewart

EDITOR
Patrick Di Justo

PUBLICIST
Gretchen Giles

MAKER FAIRE

PRODUCER
Louise Glasgow

PROGRAM DIRECTOR
Sabrina Merlo

MARKETING & PR
Bridgette Vanderlaan

CLIENT SERVICES MANAGER
Emily McGrath

COMMERCE

PRODUCTION AND LOGISTICS MANAGER
Rob Bullington

PUBLISHED BY

MAKER MEDIA, INC.
Dale Dougherty

Copyright © 2017 Maker Media, Inc. All rights reserved. Reproduction without permission is prohibited. Printed in the USA by Schumann Printers, Inc.

Comments may be sent to: editor@makezine.com

Visit us online: makezine.com

Follow us:
- @make @makerfaire @makershed
- google.com/+make
- makemagazine
- makemagazine
- makemagazine
- twitch.tv/make
- makemagazine

Manage your account online, including change of address:
makezine.com/account
866-289-8847 toll-free in U.S. and Canada
818-487-2037,
5 a.m.–5 p.m., PST
cs@readerservices
makezine.com

Issue No. 57, June/July 2017. *Make:* (ISSN 1556-2336) is published bimonthly by Maker Media, Inc. in the months of January, March, May, July, September, and November. Maker Media is located at 1700 Montgomery Street, Suite 240, San Francisco, CA 94111. SUBSCRIPTIONS: Send all subscription requests to *Make:*, P.O. Box 17046, North Hollywood, CA 91615-9588 or subscribe online at makezine.com/offer or via phone at (866) 289-8847 (U.S. and Canada); all other countries call (818) 487-2037. Subscriptions are available for $34.99 for 1 year (6 issues) in the United States; in Canada: $39.99 USD; all other countries: $50.09 USD. Periodicals Postage Paid at San Francisco, CA, and at additional mailing offices. POSTMASTER: Send address changes to *Make:*, P.O. Box 17046, North Hollywood, CA 91615-9588. Canada Post Publications Mail Agreement Number 41129568. CANADA POSTMASTER: Send address changes to: Maker Media, PO Box 456, Niagara Falls, ON L2E 6V2

CONTRIBUTORS

If you could add artificial intelligence to any household item, what would it be and why?

Linn from Darbin Orvar
Corvallis, Oregon
(Traveler's Notebook)

I want my home AI to scour the web with perfect judgment, and provide me with a summary of new and exciting information I need to improve my work.

Gordon Williams
Oxford, UK
(Open for Business)

I'd like a clothes iron that would automatically iron my clothes. It's not an obvious use of AI, but it would save a lot of people a lot of time!

Josh Williams
Ann Arbor, Michigan
(Camera-Mounted "Pinoculars")

I would add AI to the trash cans/recycling bins to keep me posted on how much I waste/recycle compared to my neighbors, while also providing ways to be more resourceful with my waste.

Sam Brown
Oakland, California
(Cat Activated Laser)

The bedroom blinds! They should open during shallow sleep: Not deep or REM sleep. Earlier on weekdays, later on weekends, and earlier if I've got an early calendar appointment.

Jordan Ramee
Fremont, California
(Editorial Intern)

Television. I'd always have someone to talk to while watching something, they'd be able to record shows for me, and they could browse my channels in seconds and recommend something.

PLEASE NOTE: Technology, the laws, and limitations imposed by manufacturers and content owners are constantly changing. Thus, some of the projects described may not work, may be inconsistent with current laws or user agreements, or may damage or adversely affect some equipment. Your safety is your own responsibility, including proper use of equipment and safety gear, and determining whether you have adequate skill and experience. Power tools, electricity, and other resources used for these projects are dangerous, unless used properly and with adequate precautions, including safety gear. Some illustrative photos do not depict safety precautions or equipment, in order to show the project steps more clearly. These projects are not intended for use by children. Use of the instructions and suggestions in *Make:* is at your own risk. Maker Media, Inc., disclaims all responsibility for any resulting damage, injury, or expense. It is your responsibility to make sure that your activities comply with applicable laws, including copyright.

POWERED BY STEAM

"When children make, even with the simplest materials, they get to realize with their hands what they can imagine in their minds. This book inspires young makers to discover creative, playful projects."

—DALE DOUGHERTY, FOUNDER OF MAKER FAIRE

Maker Camp

made possible with the support of

Make: + DK

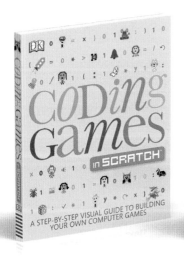

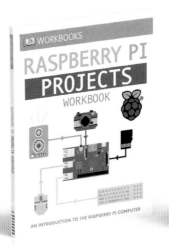
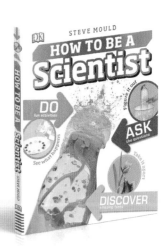

A WORLD OF IDEAS:
SEE ALL THERE IS TO KNOW

www.dk.com

READER INPUT

Unicycle Daredevils & Getting Kicked Out of Prison

OFF TO THE RACES
["3D Print a High-Power Electric Unicycle," Volume 54, page 62] has to be easily the best project I have ever done. I am 42 years old and not quite as nimble as I once was, but within about 20 minutes I got the hang of it. Man, that thing is fast! My 14-year-old son and I have had a blast on this project. Thanks a lot for all of your time and effort!
—Todd Biggs, via the web

Emilia Penttilä, Dave Darling, Phares

Author Matias Eertola Responds:
Thanks for your feedback, it's great to hear everything worked out so well with your project. Riding electric unicycles is so much fun once you get the hang of it, which in fact is much easier than most people seem to think. I wish you great times riding, stay safe! Be sure to check out Thingiverse every now and then for add-ons and future upgrades.

ART LOVE
Hey @andrewjnilsen, my son is enamored with your illustration in the latest @make magazine ["Crowdfunding Cheatsheet," Volume 56, page 16] — it has really captured his imagination. Thanks!
—Dave Darling, Dryden, Ontario, Canada via Twitter

BUILDING YOUR PROJECT AND EATING IT, TOO
I tried the Cold-Oil Spherification Bruschetta ("Culinary Chemistry," Volume 56, page 78). It turned out much better than I thought it would! Making the balsamic vinegar "spheres" was insanely satisfying! Like a controlled lava lamp or something. I'm not really a vinegar person, but they were delicious! Lots of fun to make and eat.
—Phares (15), Parkesburg, Pennsylvania

Prison Ban of the Month

» **LOCATION:** Walton Correctional Institution, Defuniak Springs, Florida
» **TITLE:** *Encyclopedia of Electronic Components, Vol. 3* by Charles Platt
» **REASON:** "Describes magnetic sensors, which are commonly used in alarm systems which could thwart security measures such as the fence alarms."

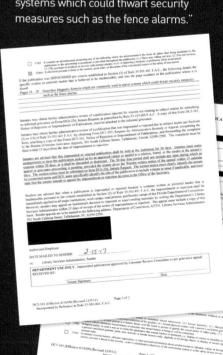

We Are All Makers

With a Handibot, the only limitation is
your imagination

The Handibot® Smart Power Tool adds a whole new dimension to reclaimed wood projects. You can incorporate your material's individual features into your design and cut them with no problem to make your piece truly stand out.

ShopBot's Richard Hill experiments with all sorts of interesting CNC designs and projects. These pieces—ranging from intricately-carved patterns to small boxes and display items—are made entirely from reclaimed wood, meaning no two are ever alike. In dealing with materials that can be different with every new job, it takes a tool with equal flexibility to adjust on-the-fly.

Thanks to the adaptability of the Handibot® Smart Power Tool: *Adventure Edition,* cutting uncommon projects is no problem. With the combination of power, precision, and portability, you won't find a more versatile tool for working in almost any material—including reclaimed wood.

Turn a distinctive feature into a defining statement with Handibot.

For more projects, posts, and videos, visit the Handibot blog: handibot.com/blog
For full tool specs and to purchase, visit www.handibot.com

www.handibot.com

open source hardware

WELCOME

makezine.com/57

Overboard

BY MIKE SENESE, *executive editor of* Make: *magazine*

CHANCES ARE IF THERE'S A BOARD ON KICKSTARTER, I'VE BACKED IT. They tempt me with their promises of project potential — I want to have them all. Sometimes I back a basic early bird model; other times I'll opt for the fully equipped top-of-the-line option. I've even justified a sub-$10 price to allow myself to buy five units of a board I had vague plans for. I keep my collection, Kickstarted and otherwise, under my TV in a wicker box, an attempt to tidy my hoarding tendencies. My hope is that if they're close by, I can tinker while I'm watching a weekend movie.

But I usually don't tinker. In fact, I've noticed that I apply the same collector mindset with these boards that I had as a kid with my action figures — most haven't been taken out of their packaging, and some are even still sealed in their shipping envelopes. I want to put them on display, and sometimes I do, periodically leaving one on my desk to stare at and study, like a map of a tiny city with roads and buildings made from miniscule components. I've plugged a few in, gotten some LEDs to blink, and even gone as far as to install Android on one. And then I carefully repackage them, first in their static bags and then into their boxes.

I'd say you shouldn't follow my lead; the wise refrain is to buy the tools you need for the project you have, not to buy a tool in hopes to find a project for it. But I'm hardly alone here. As we started working on this issue of *Make:*, I tweeted about my board-collecting habit, and got a lot of responses from others with the same tendencies. That tweet quickly became one of my most engaged, with people listing their own collections of unused boards, describing their drawers full of them, and explaining the rationale for their buying choices. (Feel free to share yours: makezine.com/go/board-hoarding)

I really don't mind this odd hobby of mine. Board prices aren't bad, my purchases support the maker community, and eventually, I'll have the perfect tool for a project that I'm sure to come across. Maybe it'll be an automated Halloween prop, or a magical winter wreath — either way, when I find it, the right board will be waiting for me, pristine in its box under my TV, next to dozens of others.

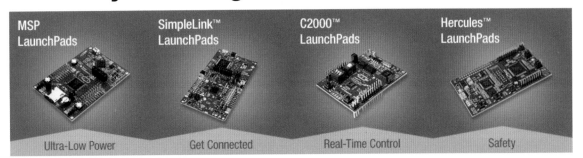

MADE ON EARTH

Backyard builds from around the globe

Know a project that would be perfect for Made on Earth?
Let us know: *makezine.com/contribute*

BOUNDLESS BEAUTY

JACQUELINERUSHLEE.COM

A palimpsest is a physical page that is reused for a new document, where traces of the original text are still visible. Historically, these were made due to lack of resources, but Hawaii-based sculptor **Jacqueline Rush Lee** has been working in this vein for the last 18 years, disassembling and restructuring books into new forms that are both surreal and organic.

Her pieces, while still clearly composed of books, evoke other familiar images inspired by philosophy and nature: a cross section of a living cell, a peacock feather, a roll of sushi, a decomposing log, a smoldering fire. "Aesthetically," she says, "I am inspired by objects that compel one to touch or be immersed within in order to ponder their significance and presence. I am also drawn to the poetry of objects and places that are at a point of decay or break down as remnants of their historical past and presence."

She manipulates the books by rolling up individual sheets of paper and bending larger sections of pages, dying different parts, and firing the books in a kiln without the use of a clay or slip.

"I'm inspired by timeless conflicts and concerns of the human condition," says Rush Lee, "Particularly, how it is as the role of the artist to try to uncover and reveal that which is unexplainable, unknowable, or at the edge of perception."

Rush Lee's work will be featured in a solo show at the Honolulu Museum of Art in the fall of 2017. —*Sophia Smith*

Jacqueline Rush Lee

MADE ON EARTH

CREATIVE CURRENCY

INSTAGRAM.COM/SHAUN_HUGHES_ENGRAVING_UK

Next time you have a pocket full of loose change, take a few moments to inspect your coins. You may be holding onto a work of art.

Shaun Hughes engraves coins with intricate and beautiful designs. This process of altering a coin with a new look isn't exactly novel — people have been doing it for as long as coins have existed and called them "Love Tokens," or in some cases "Hobo Nickels."

Hughes took a path that is counter to the direction many travel: He started with a computerized engraver and found it to be too limiting. He quickly moved to hand engraving the fine details and interesting designs of his coins, and even began hand building his own carving tools.

His designs are sometimes whimsical, completely changing the image on the coin. Other times he does intricate patterns and scroll engraving within the boundaries of the existing designs. No matter what approach he takes, his (and other artists') coins are highly sought after.

"My first coin sold for $3. My record so far is $400 for a buffalo nickel with Groucho Marx, but some by much better carvers sell for thousands," he says.

You don't necessarily have to purchase one though. Hughes has made a habit of sneaking some back into circulation: "I've put several hundred counterstamped and a few engraved coins out in change for folks to find," he says.
—*Caleb Kraft*

Shaun Hughes

MADE ON EARTH

BALANCING ACT

DANGRAYBER.COM/PORTFOLIO

Artist **Dan Grayber** combines elements from industrial design and nature to create objects that seem to defy gravity. Suspended within glass domes, large rocks are held in place with metal scaffolding, or alternatively, lend the sculpture balance by hanging motionless from wires and pulleys.

Each sculpture in his *Cavity Mechanism* series is designed around a self-resolving problem. "The idea ... came about from my thinking about inventions, and objects that exist in our everyday life," Grayber explains. "These objects all around us are all resolutions to pre-existing problems. They are designed to address a specific need that predates, and exists apart from, the invention. I was interested in creating very purposeful objects that, basically, solved their own problems — they hold themselves up."

Sometimes the dome itself inspires the work, other times he'll have a particular concept in mind that he's interested in exploring. From there, Grayber develops the idea in his sketchbook using pencil, rulers, and a compass (though he'll use CAD software to work on more complex geometries).

Grayber makes every part of the sculpture, except the fasteners and springs. Over the years he's begun using smaller materials, though he's quick to point out, "[making] smaller work is definitely not easier than larger work, it's just easier to lift."

—Lisa Martin

HEXIWEAR

The complete IoT Development Solution

www.hexiwear.com

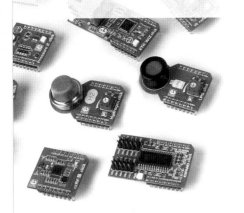
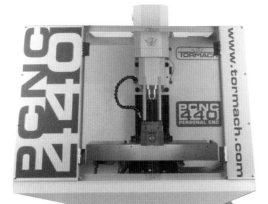

enhanced by NXP

MikroElektronika
DEVELOPMENT TOOLS | COMPILERS | BOOKS

---

smallmachine
BIGRESULTS

- x Make Precision Metal Projects at Home

- x Create Parts on a CNC Mill

- x 120 VAC - Plug it in Basements, Garages

CNC Mill Starting at $4950

TORMACH
TORMACH.COM

Meet us in the **Expo Hall**

Maker Faire
BAY AREA
May 19–21
2017

MADE ON EARTH

SOCIAL BUTTERFLIES ISTHIS.GD or VIMEO.COM/97190435 (video)

Bugs are fascinating. So ubiquitous and industrious, and so necessary on a basic level. Electronics are sort of like that, too. So when British cellular network O2 commissioned London-based artist and geek collective **Is This Good?** to promote an electronics-recycling program, the team decided to take an entomological approach.

Marek Bereza, Chris Cairns, Matt Holloway, and Neil Mendoza are the makers behind Is This Good?, who built this interactive cellphone butterfly exhibit with the help of Dave Cranmer, Robin Jackson, Liat Wasserstrohm, Nadia Oh, Luke George, Jamie Durand, and Justin Pentecost.

The butterflies, created with recycled phone parts and glittering with SIM card wings, can receive phone calls. Bereza, the software lead, explains that your phone number goes to a Raspberry Pi inside each butterfly plinth to generate art as unique to them as your phone is to you. The first digit determines color palette, the second determines speed, and so on. Bereza and Mendoza wrote in C++ with openFrameworks, which is designed for generative artwork and creative coding. "The great thing about openFrameworks," says Bereza, "is that it's very cross-platform, so the phones (Android and iOS) and the Raspberry Pi servers could all be written in the same way."

Each Raspberry Pi controls four phones and animates motors and lights, keeping everything in sync. Bereza says the synchronization was done with "an extremely hacked up version of Most Pixels Ever library." Along with the Pi, Arduinos with stepper drivers and transistors and other small electronics aid in bringing the bugs to life.

The team had to work on a deadline to build something robust enough to travel to various exhibits. However, Cairns points out that, "it was, of course, tremendously important to attach servo-controlled lasers to some of the butterflies. That real butterflies don't have this feature seems like a terrible oversight." —*Sophia Smith*

Michael Greensmith

A CITIZEN SCIENCE PROJECT
Eclipse Megamovie
https://eclipsemega.movie

A total solar eclipse is happening on Aug 21, 2017 across the United States. We need photographers to contribute so we can make a movie to study the sun! Learn more and sign up on our website.

Brought to you by Berkeley (UNIVERSITY OF CALIFORNIA) and Google

Microsoft

Make. Invent. Do.

Join Microsoft at Maker Faire Bay Area!

Come see us at our booth in the Expo Hall, for a variety of hands-on projects and demos:
Making + Coding for students | Robotics and drones | HoloLens | Mixed Reality

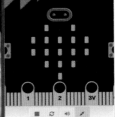

FEATURES | Commercial Drone Pilot Regulations

Clearing the Air

The FAA's drone regulations could hamper your flying endeavors. Here's how.

Written by **FORREST M. MIMS III**

FORREST M. MIMS III (forrestmims.org), an amateur scientist and Rolex Award winner, was named by *Discover* magazine as one of the "50 Best Brains in Science." His books have sold more than seven million copies.

Well-intentioned laws can sometimes set technology back by decades. Imagine life today if the Locomotives on Highways Act was still enforced. This 1865 British law required a motorized vehicle to be preceded by a man on foot holding a red flag to warn the public. The vehicle's speed was limited to 4mph on highways and 2mph in villages and towns. Weight and size limits were also provided, as were fines for violating the law.

If this sounds familiar, you must be an unmanned aerial system (UAS or drone) pilot, because some of the newly imparted rules we have to follow are reminiscent of this infamous law that slowed technological progress. For example, if a drone weighs more than 0.55lbs, it must be registered with the **Federal Aviation Administration (FAA)**. That's reasonable. What's not is that pilots who fail to register are subject to civil penalties of up to $27,500 and criminal penalties of up to $250,000 and/or imprisonment for up to three years!

HOBBYISTS VERSUS PROFESSIONALS

The registration requirement tops two sets of overlapping regulations that drone flyers must obey. One set is called Part 101, and it applies to aircraft flown for hobby or recreational purposes under the **Special Rule for Model Aircraft** (Public Law 112-95 Section 336), which prohibits the FAA from regulating hobbyists and recreational UAS pilots like me. The other, **Part 107**, applies to pilots who fly their drones for work, such as cell tower/power line inspections or photography of property by real estate agents.

The Special Rule for Model Aircraft provides that:
- The aircraft must be flown "... strictly for hobby or recreational use."
- The aircraft must be flown "... in accordance with a community-based set of safety guidelines and within the programming of a nationwide community-based organization." The leading such organization is the **Academy of Model Aeronautics**. (I'm a member.)
- The aircraft must not weigh more than 55lbs. (The vast majority of quadcopters weigh under several pounds.)
- The aircraft must be flown "... within visual line of sight of the person operating the aircraft."
- The aircraft must be flown so that it does not interfere with or give way to any manned aircraft.
- When the flight is within 5 miles of an airport, the pilot must provide "... the airport operator and the airport air traffic control tower (when an air traffic facility is located at the airport) with prior notice."
- The law goes on to provide that pilots "... flying from a permanent location within 5 miles of an airport should establish a mutually agreed-upon operating procedure with the airport operator and the airport air traffic control tower (when an air traffic facility is located at the airport)."

> Commercial flyers must apply for a waiver before flying within the controlled airspace around most airports.

In addition, the FAA has added some **basic safety requirements** to these rules for both hobby and commercial UAS pilots:
- Fly at or below 400'
- Keep your UAS within sight
- Never fly near other aircraft, especially near airports
- Never fly over groups of people
- Never fly over stadiums or sports events
- Never fly near emergency response efforts such as fires
- Never fly under the influence
- Be aware of airspace requirements

According to the **Aircraft Owners and Pilots Association (AOPA)**, "About 70% of the U.S. population lives within 20 miles of one of the 30 major airports. With over 13,000 airports in the United States today, there is a very good chance you live or work within 5 miles of one of these smaller airports." I've learned that establishing contact with airports is both straightforward and informative, and I've reached agreements with three airports within 5 miles of my flying field. (A fourth is a helicopter pad with a disconnected phone.)

The FAA's **B4UFLY app** for iOS and Android gives the names of airports within 5 miles of your location but does not provide contact information. Fortunately, the AOPA website provides phone numbers and addresses.

CERTIFICATION AND WAIVER REQUIREMENTS

This brings us to Part 107's rules for commercial pilots. If you want to sell a photo or video taken by your drone, it's necessary to first pay a fee of $150 or more and then take a 60-question test to acquire a Part 107 certificate. The Part 107 test has been strongly criticized for including technical questions that apply to manned aircraft but have no connection with unmanned aircraft.

Another major problem is Part 107's controlled airspace rule. Although a hobby flyer can simply make a phone call or two before flying within 5 miles of an airport, commercial flyers must apply for a waiver before flying within the controlled airspace around most airports. The instructions for this lengthy application form state that, "The FAA will strive to complete review and adjudication of waivers and airspace authorizations within 90 days ..." When I asked the FAA if this rule is for real, they replied by quoting the rule.

An unpredictable delay of up to 90 days makes commercial flights impossible when a client or the news media urgently needs photos of something located within controlled airspace. Although a Part 101 pilot can simply call the airport tower, the law blocks hobby pilots from selling their photos.

Provisions such as these — along with the FAA's slow response to change and the broken links on its website — lend credence to the increasingly widespread view that the FAA is suppressing drones. And its interpretation of the federal statute may even violate the First Amendment to the U.S. Constitution, which prohibits Congress from making a law that abridges freedom of the press.

PRESENTING A FIX

I propose that Congress stop this overregulation — read the specifics at makezine.com/go/faa-drone-law. For more about drones and the law, visit the FAA and Academy of Model Aeronautics online.

And please protect the reputation of UAS pilots by flying safely; don't post unlawful and dangerous drone flight videos.

FEATURES

1. *Zombie Crawler*

2. *Sand Garden* uses an Xbox Kinect, projection mapping, and a real sandbox.

3. *SpaceBox*

4. The control for *Shape Fitter* is a large coil with a gyroscope and tensegrity sensors.

5. *Orpheus Quest* is a twist on *Guitar Hero* — with lasers.

6. *Close the Leaks* is a 4-player challenge to use real air flow to direct your in-game ship.

7. *Cylindrus* is a bit like *Space Invaders*, but uses LED strips instead of a screen.

8. The handheld lantern-turned-projector the user holds to play *Fear Sphere*.

9. *Super Furry Neon Cat Heads* is like *Dance Dance Revolution* meets '90s rave.

10. *Objects in Space* puts you at the helm of a realistic control panel.

11. *Emotional Fugitive Detector* is a 2-player facial recognition game where you try to exchange undetected expressions.

TYLER WINEGARNER is the senior video producer for *Make:*, also tinkerer, motorcyclist, gamer. Reads the comments. Uses tools, tells stories. Probably a human. Tweets @photoresistor

We Are All Makers

GAME CHANGER

Written by
TYLER WINEGARNER

alt.ctrl.GDC showcases wild and weird control schemes that create a whole new kind of immersion

You're in the basement of an office complex. The power is out. Your dim lantern lights the way. You don't know how to escape, and you're not alone — there's some sort of monster down there with you. You shine the beam of light over a sign that reads "Lab SB-132" and a voice crackles over your headset: "Wait, stop there. Go down the hallway to your right and take the last left." You head down the hallway, but in your disorientation, you can't remember if she said the last left or the first. The growls are getting closer.

The game is called **Fear Sphere**, played entirely using the lantern in your hand. The beam of light is created by a pico projector housed in the lantern body, and its orientation is determined by a six-axis accelerometer. The only thing you can see is the small circular projection onto the inside of a pitch-black sphere, like a flashlight in a dark room. You can call out your location to a friend outside, who has a map and can help direct you, but it's easy to get panicked.

MATTER OVER MIND

When you think of a game controller, you're likely picturing a shapely lump of plastic, with two easily graspable lobes, two joysticks, and as many buttons as can be placed within reach. You probably don't think of a pile of sand, or a cardboard box, or an endless loop of carpet on a treadmill roller.

alt.ctrl.GDC is a showcase at the annual Game Developer's Conference that explores alternative interfaces for interacting with games. It started as an award category in the Independent Games Festival, but as entrants grew, it became clear that these unique games needed a home of their own. Shepherded by John Polson, alt.ctrl.GDC is now in its fourth year, showcasing custom-built hardware and software to create unique player experiences. "We spend a lot of time thinking about how video games touch us," says Polson, "alt.ctrl explores how we touch games."

IMAGINATIVE ENGINEERING

Unsurprisingly, microcontrollers like Arduino and its endless variants are the brains behind many of these interfaces. Numerous commercial game engines, like Unity and Unreal, provide an API that allows Arduino to interface directly with the engine as an input/output device. Events in the game can trigger anything from an indicator light or movement in a physical gauge to kinetic interactions like activating fans or blaring klaxons. And for input, the only limit is whatever the creator can imagine and engineer.

Take **SpaceBox**, a playful and childlike spaceflight simulator that you control with a simple cardboard box. You lean forward, back, left, and right to steer the ship. Designer Robin Shafto wanted to recreate the imaginative worlds of Spaceman Spiff, Calvin's spacefaring alter-ego from the *Calvin and Hobbes* comics. Buttons mounted underneath the box detect the direction you're leaning. To raise your shields, you lift up the left and right flaps of the box lid. These movements are detected by accelerometers, and the ship's blaster is a simple conductive button built from a lip balm tin. The colander you wear as a helmet might just be for looks, but the experience wouldn't be complete without it.

Not every solution needs to be high tech. **Zombie Crawler** lets the gamer play a zombie crawling down a hallway to reach the tasty human at the end. To do this, players drag toward them a loop of carpet on treadmill rollers, occasionally slapping at green buttons on either side to smash obstacles out of the way. When the human takes aim with the shotgun, the player can rock the treadmill from left to right to dodge. The movement of the carpet is captured by a computer mouse's optical sensor, and the tilting movement is recorded using magnetic reed switches. All of these inputs are captured using an IPAC board, a computer interface card popular with home arcade cabinet builders, which is an easy solution for folks who want to spend more time on their hardware design than their Arduino code.

ENHANCING THE EXPERIENCE

While some projects are created just for the experimental whims of the teams behind them, others come from universities, like the Game Design and Graphics program of Uppsala University in Uppsala, Sweden, where alternative control schemes are part of the educational curriculum. There's an annual alt.ctrl game jam with gathering sites in 11 cities around the world. Some creators, like Gregory Kogos (of **RotoRing**, a simple interface with two Adafruit NeoPixel rings and an Arduino) are developing their alternative control games with the intent to launch a commercial product. Likewise, **Objects in Space** is a trading game augmented by an open source control interface powered by Arduino. You can play the game without the extra control surfaces, but let's be honest: launching torpedoes with a toggle switch is always better than a mouse click.

> "the only limit is whatever the creator can imagine and engineer"

FEATURES | *Maker ProFile*

Stepping Up

Ran Ma is developing **smart socks** to help monitor **diabetic health**

The battery on the market-ready model is machine washable, and stays affixed so that it doesn't get lost.

The first two prototypes had visible conductive thread, and unwieldy external electronics.

The third prototype featured a snap-on case.

22 We Are All Makers

With a $50,000 grand prize, the TechCrunch Hardware Battlefield is one of the high-profile events at the annual CES conference in Las Vegas. This year's winner was Siren Care, a wearables company that embeds electronic sensors into clothing in order to track changes in a person's health. The company's first product is smart socks that people living with diabetes wear to monitor temperature and detect injury in real time, since they often have nerve damage and are unable to feel pain. Ran Ma, CEO of Siren Care, was previously the CMO of Voltset, a Danish consumer electronics startup. She successfully ran the Voltset Kickstarter campaign, which raised $113,548 (192% of their goal).

Q. You launched Siren Care into a category that already had some strong products. Did that worry you?
A. We do have competitors, which is great. If no one else is doing something similar, you're either a crazy genius or you're just crazy. So having competitors, that's good. It tells me that I'm not completely crazy.

In addition, we are thinking long term. Ideally we are developing technology that can generate a family of products, using the same personnel, the same expertise, and the same intellectual property.

Q. What *is* your long-term vision?
A. We're about using cutting-edge smart fabrics to make products focused on health and prevention, about using different techniques to integrate electronics into fabrics.

We're focused on taking wearables to the next level. We want them to be discrete, continuous, and we don't want you to have to make a behavior change. We want them to fit into your life, so our socks are machine washable, machine dryable, and you don't have to charge them. They are just like normal socks. That's the next stage of wearables.

Q. What lessons did you learn from your first product, the Voltset multimeter?
A. It opened my eyes to the possibilities of the Internet of Things and connected devices. Voltset really helped me learn about prototyping. It taught me that if you have an idea, you can make it.

Another key lesson I learned: build a strong team. You'll depend on them for solving problems. Once I knew the product I wanted to build, the next important step was to build a world-class team with expertise in the fields I was working in: textiles and electronics. Ultimately you want to have a team that can help you manufacture and ship hundreds of thousands of units.

No matter your skill level, or engineering level, there are so many resources out there. The first prototypes may not be pretty, but that experience is going to help you. Making those first prototypes was really important because I now understand the product on a fundamental level, because I made it with my own hands.

Q. Tell me about your experience making prototypes.
A. I hand-sewed prototypes in my room. I bought conductive thread from SparkFun, and used an old Arduino. I took a soldering class at a hackerspace in San Francisco. I went in there and was like, "Can someone help me solder my sock?"

The first couple were scary: wires coming off everything, the solder was terrible. I'd go to medical conferences to show it to doctors and they weren't impressed. But I just kept working at it. The electronics got smaller, my app got better, the socks got smoother. Eventually the medical community started to buy into it.

Q. It sounds like you had a lot of setbacks. What kept you going?
A. Going out and talking to potential customers, potential patients. We talked to nurses, care homes, doctors, wound experts. I went to a lot of conferences. I talked to diabetic patients about the product, and they would get emotional. People were so afraid of losing a toe or getting a foot amputated, losing their independence. Every time I felt like giving up, I just remembered: this is important.

Q. Do you think the game has changed now that sensors and prototyping tools are more accessible?
A. Yes. Sensors are getting smaller and cheaper every day, which is great for wearables and IoT devices. It allows more people to start making prototypes. That's very important because when you have an idea that you're having trouble explaining, you can get it into physical form and show it to people.

Q. You used crowdfunding for the Voltset, but went after venture capital for your second product. Why the change?
A. It's different for every company. It's important to think about what you want to build. Some of my friends' companies are fine just coming off Kickstarter or Indiegogo, or they can survive on revenue. Other companies go the VC route because they need a lot of capital for R&D or manufacturing. We wanted to build a large company. We've spent a lot of time in R&D, figuring out technology and a road map for the future. So in our case, fundraising has been a priority, because we want to take on outside capital to grow faster.

Q. You're launching healthcare products. Was the prospect of dealing with the FDA and other regulatory agencies intimidating?
A. Going the healthcare/regulator route can be challenging and complex, but there's also some protection. If you do some research and consult with experts, you can unravel some of the regulations. Many FDA clearances aren't that hard to get. Once you get that clearance, it's good for you: you've built a moat around yourself. That means it's going to be that much harder for people to copy you.

DC DENISON is the co-editor of the *Maker Pro Newsletter*, which covers the intersection of makers and business, and is the senior editor, technology at Acquia.

Visit makezine.com/go/siren-care to read the full interview.

Special Section | **Board Guide 2017** | The State of Boards

Put a Board on it

PLATFORMS, PRODUCTS, AND PURPOSES: THE CURRENT CROP OF MICROCONTROLLERS VIES FOR YOUR ATTENTION

Written by Alasdair Allan

ALASDAIR ALLAN is a scientist, author, maker, and journalist. Currently freelance, building, breaking, and writing.

THE STATE OF BOARDS

OVER THE LAST FEW YEARS, WE'VE SEEN A HUGE GROWTH IN THE NUMBER AND VARIETY OF BOTH MICROCONTROLLER BOARDS AND SINGLE-BOARD COMPUTERS. Just as in the early 1980s, when the arrival of cheap home computers led to an explosion of variety and choice, the growth in the number of microcontroller boards in today's market has meant that manufacturers have experimented with both features and form factors.

However, we live in a different time, and the trends driving the growth in microcontrollers have led us down a different path than they did last time around. In the 1980s, we looked at the new home computers and saw not just glowing screens, but boxes that could be manipulated in any number of ways. Today's computers, smartphones, and tablets instead are seen as a way to communicate. Nowadays a microcontroller — or even a "real" computer — without an internet connection is just a brick.

LOOKING BACK
Today's microcontroller board market began with development boards. Essentially, these were breakout boards for new chips that manufacturers wanted to bring to market. They allowed professional engineers to experiment before they placed orders of thousands, or perhaps millions, of chips to put inside their products.

From a hobbyist perspective, these development boards were built for professionals and were generally too expensive to be useful. For the most part, the now venerable PIC microcontroller was the backbone of the maker movement's electronic builds and came by the chip, rather than by the board.

The modern era, defined by microcontrollers becoming conveniently packaged on boards, began with the Arduino. The "little blue board that could" has changed the way that we do electronics, not just for hobbyists, but for professionals as well. Those expensive — and badly documented — developer boards for the professional market have given way to cheaper microcontroller boards that are far more easily accessible. That's been good for everyone, including the professionals, and we have makers to thank for that.

LOOKING FORWARD
It's safe to say that the growing popularity of internet-connected smart devices, the so-called Internet of Things (IoT), has changed the face of the microcontroller board market.

The current generation of boards now come with radios, sometimes lots of radios. Before the IoT, microcontrollers, like computers in the past, were seen as a way to automate, or control. Now, they too have become communication tools.

It's just that, for the most part, they're talking to each other, rather than to us.

THROWING IN THE KITCHEN SINK
The microcontroller board market is in transition. Just as the way we're using computers is changing, the way we build

Special Section | Board Guide 2017 | The State of Boards

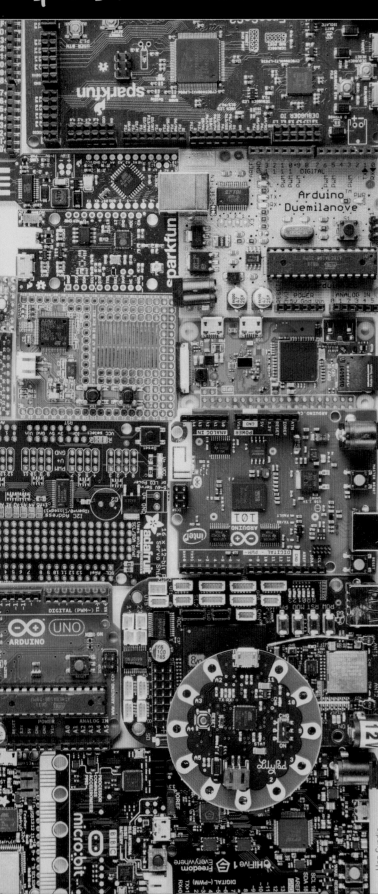

hardware is changing with it. Because of that, manufacturers aren't entirely sure how people are going to use their product. The response from many has been panic, and they "throw another radio on it."

The arrival of what I call "kitchen sink" boards, which try to be all things to all people, has been one of the main trends over the last year or two. This is especially evident on Kickstarter, where people are desperately seeking to differentiate their board from all the others.

Microcontrollers are ultimately used to control things, and that means there isn't a single-use case. But that doesn't mean it's a good idea to have one board — with all the power and all the necessary radios — to do all the jobs a microcontroller might be asked to do. A typical kitchen sink board comes with multiple radios, and more CPU and RAM than most embedded devices will ever need to do their jobs. And this hardware is expensive. "One board to rule them all" will never be the right board to use. As with the UNIX command line, people should try and focus on building small, simple hardware tools, not giant monoliths.

FORM MATTERS

One casualty that arose from the end of early home computers is now evident: the decreasing number and variety of form factors in which those computers came. We're now seeing the same thing for microcontrollers and, to some degree, single-board computers.

The "classic" Arduino layout, including the irritating, irregular offset between pins 7 and 8, has become a standard, almost by default. In addition to clones and imitators, the huge community surrounding the board has brought with it shields and other hardware designed for its configuration. This means boards that might not resemble an Arduino computationally still resemble it physically.

Other board makers have started to see their design become standardized now too. For instance, Adafruit's range of Feather boards has a standard layout, one that imitators and competitors are starting to duplicate.

There's also a movement occurring at the smaller end of the market in which manufacturers have started to produce integrated modules on a single board. Often destined to be mounted on other circuit boards, the castellated module is now a default way to get today's tiny surface-mount parts into the hands of a wider community that often doesn't have the tools, or the skills, to make use of them directly. This became especially obvious

People should focus on building small, simple hardware tools, not giant monoliths.

with the arrival of the ESP8266, which led to an ESP-12-like form becoming the default. Competitors like the RTL8710 now come in very similar configurations. Some are even pin compatible.

Similarly, the Raspberry Pi's layout has been imitated, with several newer boards duplicating it exactly. One of those, Asus' Tinker, is rapidly carving out a niche as an inexpensive media center. And the popularity of the Raspberry Pi Zero, along with the recent arrival of a wireless variant that has made the board far more useful, may start to drive imitators. But we aren't witnessing a full form factor standardization for single board computers — at least not yet. Like the Arduino's pin headers, the Raspberry Pi's header block has become a standard by default, and for the SBC market, perhaps that's enough.

COMPUTING THAT IS CHEAP ENOUGH TO THROW AWAY

General-use microcontroller boards with onboard Wi-Fi can now be found for less than two dollars, while a single-board computer can be picked up for only a few dollars more. Even for those of us that have grown up with Moore's Law, that seems almost inconceivable. And yet, we're getting to a place where computing is not just cheap, it's essentially free.

That changes how people are using microcontrollers. The ESP8266 has been a runaway success, and in many ways is the opposite of the "kitchen sink" boards that manufacturers — unsure of their markets — are pushing as the solution to the IoT.

"Good enough" is sometimes all that's needed.

The ESP8266 is also successful due to the community that has rapidly grown around it. This community coalesced not because of the features the board offered — there have been other small form factor wireless boards — but because of one feature the other boards didn't offer, the price point. As a result, the ESP8266 has become the "third community" of the maker electronics world alongside the Arduino and the Raspberry Pi. Although some of that success can be attributed to the ESP8266/Arduino compatibility, the chip's community-built Lua development environment is actually far more widely used, which suggests the price point really is the thing that drove community adoption. It appears that "good enough" is sometimes all that's needed.

THE ARRIVAL OF FPGA

Field-programmable gate arrays (FPGAs) are a very different kind of beast than a microcontroller. With microcontrollers, what you have control over is the software, the code living on the chip. With an FPGA, you start with a blank slate. You design the circuit. There is no processor to run software on until you design it.

It might sound crazy, but what this gives you is flexibility, and the age of the maker FPGA has arrived without much real fanfare. There is now an open-source toolchain for Lattice's iCE40 FPGA, and FPGA boards specifically targeting the maker market — like Alorium's XLR8 — are starting to appear. These boards provide hardware-level flexibility, allowing you to adapt hardware rather than replace it as your project evolves — something maker projects have a tendency to do over time.

It's also been interesting to see the appearance of FPGA-like chips inside "real" products. For instance, Apple's new AirPods are actually built around a Cypress PSoC chip.

PACKAGING MACHINE LEARNING

One of the most intriguing features of the Arduino 101 board when it was released was the 128-node neural network hidden inside the Intel Curie driving the board. For months after its release it was almost impossible to get any information about, or access to, the network, with Intel promising that documentation and library support were "coming soon." That's changed with the arrival of the CurieNeurons Library from General Vision. A free version gives limited access; the "Pro" library offers full support at a cost of $19 per user (which is almost two-thirds the cost of the board itself), and that's going to be too rich for most makers.

Other boards with machine-learning features, like Intel's Joule, are also starting to make an appearance. But, like most of Intel's offerings to the maker market, it's pitched toward the high end. In a market where low-end boards are routinely stretched to do things most people thought they couldn't do, this can be a hard sell.

WHAT DO PEOPLE REALLY WANT IN A BOARD?

Most people, and most makers, want to solve a problem. While, for some, the specifications of the board really matter, those people are by far the minority. What some manufacturers fail to understand, sometimes repeatedly, is that most people don't need more performance than what they require, and would rather pay less for the right tool than extra for something excessive. In the end, most people aren't interested in the kitchen sink, except when they need to do the washing. ●

Special Section | **Board Guide 2017** | Standouts

Extraordinary

EVERY YEAR, COMPANIES BIG AND SMALL RELEASE SCORES OF NEW MICRO-CONTROLLERS, increasing the processing power and features we have access to. Sorting through the options can be the hardest part of a project, so we've selected our hot new board highlights in four key categories: Education, Robotics, IoT, and Art. These stand out for their advanced specs, built-in offerings, and in some cases, for their innovative interface options.

Robotics Boards

Robotics is a field that encompasses a lot of different skill sets — to make a great bot you need to work with mechanical engineering, electronics, and programming. This can be quite a task for a team, let alone a single developer trying to make his or her own concept come to life. One way to make your development processes easier is to start with a board that is designed to help you build a robot. Any microcontroller can be a start, but having a board that has built-in features like motor controllers and sensors can get your bot rolling, walking, or wiggling with less development time.

Make: 2017 Board Guide STANDOUTS · Robotics

BEAGLEBONE BLUE
» Price: $82
» Size: 3.4" × 2.1"
» Type: Single Board Computer
» Clock Speed: 1GHz

BeagleBone Blue will be familiar to anyone who has used a BeagleBone Black, but offers some new functionality in order to target robotics and machine control. It provides a fully open source, real-time embedded Linux platform in a small footprint. The board has a wide array of possibilities for sensors and connectors, including support for LiPo, Wi-Fi and Bluetooth, 6V servo and DC motor outputs, along with four encoder inputs. It also has a built-in IMU and barometer, and there is USB 2.0 for client and host. On the user interface side you will find programmable LEDs and buttons. Possibilities for expansion are endless, including headers for motors, i2C and SPI busses, and more. —*Philip J. Angileri*

NVIDIA JETSON TX2
» Price: $599
» Size: 6.7" × 6.7"
» Type: Single-Board Computer
» Clock Speed: 2GHz

Nvidia's Jetson boards bring parallel processing for deep neural networks for visual recognition and other AI tasks. The TX2 is the latest board in that line. It trains and works faster, recognizing objects in an image in mere milliseconds. Despite its horsepower, the TX2 core runs on low wattage, suitable for battery-powered robots.

The TX2 comes mounted on a mini-computer-sized developer board with a camera and other goodies. Pop the module off the board to have a minimal-sized core for your custom gizmo, or keep it on the developer board to use it as a single-board computer.
—*Sam Brown*

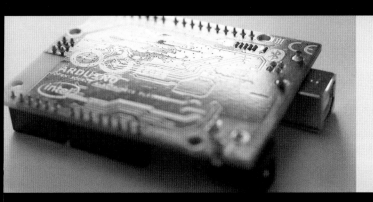
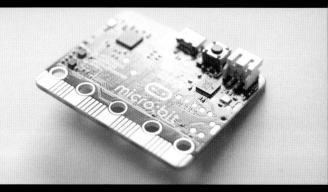

Boards

THESE EIGHT DEVICES ARE PUSHING THE LIMITS OF WHAT YOU CAN DO WITH DIY TECH

Written by Matt Stultz • Photographed by Hep Svadja

Educational Boards

While any board on the market — especially some of the venerable classics — is going to be a learning experience, these two boards have a step up on the game. When picking educational boards, we looked at the curriculum and guides available for the platforms, along with how they could be best used in an educational setting. If you are a teacher, we think these have something to offer you and your students.

Make: 2017 Board Guide STANDOUTS • Educational

ARDUINO 101
» **Price:** $30
» **Size:** 2.7"×2.1"
» **Type:** Microcontroller
» **Clock Speed:** 32MHz

The 101 keeps the same form factor as other standard Arduinos, but replaces the usual Atmel chip for the Intel Curie module — the first dev board to carry this mighty little chip. While still programmable through the Arduino environment, the Curie brings a 6-axis combination accelerometer and gyroscope along with Bluetooth Low Energy radio for wireless connectivity. For educational use, it's hard to go wrong with boards that work with all things Arduino (shields, IDE, code); the 101 will slide right in.

To help push the 101 over the edge as a great educational board, it is one of the few that is supported by Google Science Journal, letting users connect additional sensors they can log with the app. —*Matt Stultz*

BBC MICRO:BIT
» **Price:** $18
» **Size:** 1.97"×1.57"
» **Type:** Microcontroller
» **Clock Speed:** 16MHz

The micro:bit is a standout for education. First, it uses a block-style programming language, with drag-and-drop code. This is a forgiving environment to learn in. There are visual cues for how loops and conditionals fit together. And there are no persnickety semicolons or spacing to trip up the new learner. Second, it has enough lights, buttons, and sensors built in to give a student programmer plenty to play with before they have to learn how to wire in additional parts. Finally, while the micro:bit is a newer board, it already has a healthy user community and plenty of lessons. —*Sam Brown*

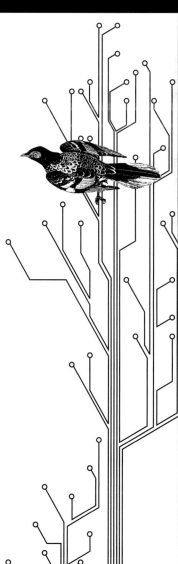

Special Section — Board Guide 2017 | Standouts

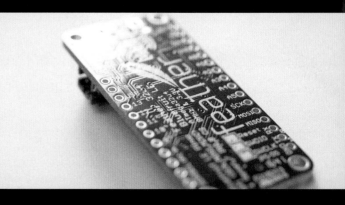

PHILIP J. ANGILERI has been an industrial designer for over 20 years. He is president and principal for design and engineering firm NarrowBase (narrowbase.com), and is an active member of Ocean State Maker Mill, where he co-founded their Robotics Club.

SAM BROWN tinkers with circuits and pretty much everything else.

IoT Boards

For better or worse, we are connecting everything to the internet. Startups are racing to connect our coffee pots, cars, doors, and even our pets. Some makers are just going to connect things to the internet for their own enjoyment. Whatever your interest, a good internet-connected board will help immensely — getting an IoT prototype to work quickly can be the key to winning a hackathon or even getting the VC money. Our IoT picks both come with wireless connectivity that will help you create your own platform or put your device on an existing one.

Make: 2017 Board Guide STANDOUTS • IoT

RASPBERRY PI ZERO W
» **Price:** $10
» **Size:** 1.18"×2.56"
» **Type:** Single-Board Computer
» **Clock Speed:** 1GHz

The Raspberry Pi foundation has brought together a robust community, sharing ideas and projects that have created a diverse knowledge base around their product. This makes it easy to create projects around the Raspi platform. The Zero W keeps the same form factor as its predecessor, the Zero, but that ending W stands for Wi-Fi. For IoT projects, the Zero W supplies you with everything you need and very little of what you don't. Although double the price of the original, non-wireless model, the Zero W's small size and still low cost make it an easy fit into any project and any budget. —*Matt Stultz*

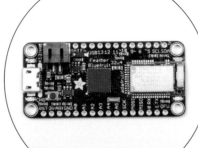

ADAFRUIT FEATHER HUZZAH
» **Price:** $17
» **Size:** 2.0"×0.9"
» **Type:** Microcontroller
» **Clock Speed:** 80MHz

The ESP8266-based Adafruit Feather Huzzah is a great way to connect your embedded project to the internet. The Huzzah comes preinstalled with the NodeMCU Lua environment, but Adafruit provides both board specification and libraries that make it easy to use with the Arduino IDE. The board is part of the Adafruit Feather lineup, so you can expand the basic board with "Wings" for LEDs, displays, relays, and motor control. The popularity of the ESP8266 has meant a lot of code is available to extend its functionality including running MicroPython, and even a way to emulate a WeMo home automation device, which means controlling your board from your Amazon Echo. —*Kelly Egan*

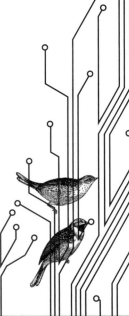

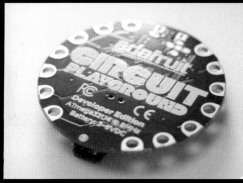

KELLY EGAN
(kellyegan.net) is an artist, teacher, and creative coder living in Providence, RI.

MATT STULTZ
is the 3D printing and digital fabrication lead for *Make:*. He is also the founder and organizer of 3DPPVD and Ocean State Maker Mill, where he spends his time tinkering in Rhode Island.

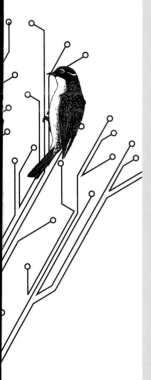

Art Boards

Wearables, art installations, and other creative projects all come with their own challenges. Boards with unique features out of the box can bring useful solutions to these types of endeavors. Boards that also offer well-documented libraries for these features save the user programming time, which can be even more important. Our art board picks have sensors and outputs that help make interesting projects easy and fun, and that make programming with them as simple as possible.

ADAFRUIT CIRCUIT PLAYGROUND
» **Price:** $20
» **Size:** 2.0" diameter
» **Type:** Microcontroller
» **Clock Speed:** 8MHz

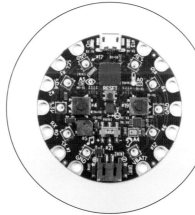

Ten multicolor LEDs and a battery connector are the top items on Adafruit's Circuit Playground's qualifications to be at the heart of your next light-up fashion piece or cosplay project. It's sewing-friendly, with a circular shape that doesn't snag on cloth and connectors you can tie conductive thread onto. Adafruit also includes an accelerometer, buzzer, switch, sensors for light and heat, and two buttons onboard. That's enough inputs to make your piece able to swap between different modes and react to the outside world, without ever having to break out the soldering iron. —*Sam Brown*

SPARKFUN MICROVIEW
» **Price:** $40
» **Size:** 1.05" × 1.04"
» **Type:** Microcontroller
» **Clock Speed:** 16MHz

Adding a display to an embedded project often requires more effort than everything else. Not so with SparkFun's MicroView which combines an Arduino-compatible board and OLED display in one small package. SparkFun's library makes drawing to the display similar to using Processing, with shapes, text, as well as slider and gauge widgets perfect for showing sensor data. It is a little over half an inch tall, which might get bulky on your wrist but would work as a tiny game platform, dollhouse TV, or even in a large necklace. One note: You'll need to buy a programmer to upload your code. —*Kelly Egan*

Special Section **Board Guide 2017** | Espruino

Open for Business

enough to pay me a salary — without the need for Kickstarter campaigns to prop me up.

PORTING PRODUCES PROBLEMS

Because the firmware and tools are open source, Espruino is being used on more and more devices. The Espruino community ported Espruino to the ESP8266, and as of the end of 2016, more people were using ESP8266 than all of the boards I sell combined. In fact, only one-third of Espruino users were actually using a board that I'd made.

This is fantastic, but it does lead to some problems. I'd always prided myself on offering really good support to users on the Espruino forums — often spending two hours or more every day helping out. As the fraction of people using my boards has dropped, I've found myself spending an increasing proportion of my time effectively working for free.

There are very clear benefits of an open source hardware business — for everyone. The costs of individual electronic components and production drop off drastically at higher quantities, meaning that most open source hardware companies can afford to sell their products for less money, while still giving you all of the designs so that you can see how they work and fiddle with them.

Open source software usually relies on offering a consulting business around an open source product — where you can work with your software more efficiently than others can. However, things get a lot murkier when open source hardware and software collide — which will be happening

The first panel of 30 Espruino Pico boards off the production line.

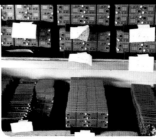

Assembled Espruino boards at Seeed, ready for programming and testing.

Servomotor control using Espruino on a touchscreen STM32 dev board.

more and more in the future.

In Espruino, the aim has always been to replace hardware complexity with software complexity. The hardware is very simple, but the software is extremely complex and takes the vast majority of the development time (at least 10 times more). I have to mark up the cost of my hardware in order to pay for the time it takes to develop the software. This makes my boards very uncompetitive compared to a $2 ESP8266 board that's had Espruino flashed onto it for free, or even another company selling a board with Espruino preinstalled.

This does hurt sales a little, but the biggest problem is actually that of support. If you have a problem with a normal open source hardware board, you'll talk to the manufacturer. However, if you have a problem with an Arduino clone that you bought from eBay, you'll go to the Arduino forums. The same is true of Espruino, and users of these other boards generally have more problems because of bad documentation and out-of-date/buggy firmware. It can

be very frustrating to provide an appropriate level of customer support *and* not spend too much time on everyone else but myself. This is especially difficult when I can't even determine who is using what board!

In normal open source software, you'd hope that contributions from the community would go a long way towards making up for this, but when writing tools for embedded software, this may not happen as much. The majority of my users like Espruino because they'd prefer not to write C code, so it's highly unlikely that they'll dive into the optimized C source code of the interpreter. When writing software for the desktop, you can write tests to ensure it works correctly — and while this is done as much as possible with Espruino, to fully test the hardware APIs, you have to run them on the hardware itself. Although most contributors test their code on their own boards, they're very unlikely to test it on the four types of Espruino boards I make, let alone the 40 other types of boards that Espruino runs on. So even

vetting contributions can take a lot of time.

To top it all off, the software that requires all of this effort to maintain has little to no value. I'm unable to license it to any interested companies, because they can use it however they want without a need for one. Instead, all I can do is charge for services I provide on top of Espruino, such as maintaining a port or adding some functionality.

Although I'm able to pay myself a salary, it's purely out of the generosity from the community, not because I have a great business model. I can't help but feel that the lack of a clear business case for the use of open software tools with open source hardware is hurting innovation in the field. I think it's telling that there are now hundreds of venture capital–funded Internet of Things websites. At the same time, despite millions of users, the Arduino IDE hasn't noticeably changed since the first alpha release 10 years ago, while the Codebender online IDE has had to shut down.

Having said all that, I am in the very fortunate position of being paid to do my own thing. Espruino wouldn't be as popular as it is if it weren't open source, and I use open source software myself almost exclusively. I've had some fantastic contributions to Espruino from the community (for example, the ESP8266 and nRF52 ports), and even bug reports from users of other boards have improved Espruino's reliability for everyone. All things considered, I definitely benefit from open source software, while feeling like I'm also giving something back. ●

Special Section | **Board Guide 2017** | Adafruit

The Ripe Stuff

WITH OPEN SOURCE IDEALS AND AN ENGINEERING GENIUS AT THE HELM, ADAFRUIT IS A MAKER REVOLUTION IN MANUFACTURING — AND SO MUCH MORE

Written by Gareth Branwyn

THERE ARE SOME COMPANIES THAT JUST SEEM CHARMED, POSSESSED OF A DEFT BALANCE OF CORPORATE INSTINCTS, WITH THE COURAGE AND CONVICTION TO LEAD, to make bold, even audacious moves that repeatedly turn out to be the right ones. Adafruit Industries is such a company. Founded in a dorm room in 2005 by MIT engineer Limor "Ladyada" Fried as an online learning resource and marketplace for do-it-yourself electronics, Adafruit is now a highly successful community-driven electronics company, educational resource, and maker community thriving in SoHo, Manhattan.

Limor sees three keys to the success of the company: "Being focused on others, having an unconditional belief that you can be both a good cause and a good company, and seeing risk-taking as your friend and your only real competition as yourself."

It is these high-minded tenets that make Adafruit something special. That they've managed to hold on to these values as they've grown — creating an increasingly open source company culture around them — is especially laudable.

Limor embraced open source before many other makers going pro did the same, largely from her environment. "Boston and MIT had a very strong open source culture," she says. "In my youth, I spent time hanging out with GNU/Free Software Foundation people and others in the open source software community. I learned to code by looking at open source software, so it only seemed natural to try and apply that same philosophy to my fledgling hardware business.

"I didn't think it would 'pay off' at all," she says. "But it was just something I felt very strongly about, and if it worked out, if it became profitable, then great."

From the very beginning, with early product successes like the MiniPOV (persistence of vision) and SpokePOV (POV for your bike wheel) kits, and the MintyBoost (the world's first open source device charger), Adafruit has always made well-designed, user-friendly kits and components accompanied by excellent online tutorials and build videos. The company now carries more than 3,500 products, including over 400 original designs. With these efforts, and educational offerings like the Adafruit Learning System, Adafruit has done more than just about any other organization to humanize electronics to make them fun, accessible, and useful to the tinkering masses.

PASSION PROJECT

Anyone who knows Limor, and her partner in crime, Phillip Torrone (aka "Mr. Ladyada"), knows that they live and breathe Adafruit. This company, this community, this mission, is a passionate 24/7 commitment. How can Limor handle being CEO of the company, the star (as Ladyada) and co-creator of so many of Adafruit's YouTube shows (and other online content), and the main product designer? It's exhausting just to think about it all.

"A lot of it is just knowing how to do triage on projects and manage my time effectively,"

Hep Svadja

Special Section | **Board Guide 2017** | Adafruit

Michele Santomauro and Vance Lewis holding component reels in preparation to load the pick and place machines.

Limor at her desk at the end of the day. This is where she broadcasts *From the Desk of Ladyada* as well as other shows.

Vance debugging chips that were rejected in testing. Debugging these chips lets the manufacturing process become more successful as well as more efficient, according to Michele.

Employees at workstations on the manufacturing floor.

The Adafruit team.

she says. "In planning what I'm going to do, I often decide based on what can get the most people on our team going. This maximizes my efforts." She continues: "When leading a group, it's important to identify what we call 'NP problems' ["Nondeterministic polynomial time"] — these are tasks that may take some time and care to complete, but they can be quickly verified. Such 'NP problems' can be given to people on the team so that they can quickly take them on, practicing and learning new skills as they tackle them. Then, we can come back together and I can check in on the final result."

Limor also maximizes her content creation time by doing lots of video live — eliminating the time-consuming editing process. "Oftentimes, I will do a video just as I'm wrapping up work on something I was doing anyway. Or I might design a PCB live, on-air. So, I'm able to do 'double duty' for that time.

"It's a balance of strict-time and free-time scheduling," she says about managing her days. "For example, the live shows happen every Wednesday night, and we have weekly group and individual meetings that can't be moved. And then there's the free time in-between. 'Free time' is for handling the little and immediate things that come up. The toughest part is looking at the free time available and figuring out the right task to fit into that time slot. It takes practice and experience, knowing what you can realistically complete within the allotted period."

Despite the success, Limor stays active with engineering. "I still design or review all of the hardware designs as well as approve the samples for other things that we stock." She adds: "KTOWN [Kevin Townsend], our amazing wireless and embedded engineer, does the firmware, design, and hardware for our more complex and valuable hardware projects." Limor spends most of her time on the testing procedure and fixture design, and says she pretty much does all of it. "I make one tester, then our in-house fab team helps build the testers from my prototype." Limor believes that having good test and quality assurance procedures are a must. She says that a significant amount of the time and cost of a product is taken up in the test/program/verify stage. "So, I work a lot on optimizing that," she says. "We reuse a lot of our designs to do so. For example, all of our dev boards are tested using a Raspberry Pi 3 running OpenOCD (an open source on-chip debugging program), with a PiTFT screen (an RPi-compatible touchscreen made by Adafruit)." She continues: "They make for lovely, small, stand-alone testers that are easy to duplicate, and they're very inexpensive."

HOW IT'S MADE

"I develop products by listening to people in the community," Limor says. "At places like Maker Faires, tech events, and on our *Show and Tell*, I see what people are working on, what projects are popular, and what problems they are having. Then I think of ways of solving them.

"Sometimes, I'll have these 'problems' stewing in my head, and whenever I see a new chip or a design idea, I'll 'test' it against one of the problems I've been thinking about," she says. "Sometimes, that means adapting an existing solution to a new use. I'll give you an example: the PCA9685 is an LED driver chip. But when I saw its datasheet, the adjustable PWM (Pulse-Width Modulation) rate and high-bit-timer meant that it would be perfect for controlling servomotors and robotics. That wasn't a recommended use on the datasheet, but it has now become the top use of the chip."

Adafruit's Circuit Playground, an all-in-one board with built in sensors and LEDs (and one of their top sellers), came about because teachers and parents needed something easier to start kids with. "[They] kept telling me that they loved wearables and Arduino, but they couldn't get a student's laptop set up, teach breadboarding, and cover the basics, all in a 45 minute class. They wanted something that required no soldering or breadboarding. And they wanted it for $20. As any engineer knows, the bigger the constraints, the more fun."

Limor is quick to point out that this is not the most time-consuming part of new product design. "Hardware is, arguably, the easiest part of the puzzle to solve. The tough parts are firmware, software, and support," she says. "For instance, right now, we're focusing a lot of our effort on MicroPython/CircuitPython. That effort is being headed by Tony DiCola and Scott Shawcroft. The hardware takes only a few days to design, but doing a good port of the Python core to the processor takes many months."

Regardless, Limor's not an insomniac. "I actually sleep for 10–12 hours a day. I sleep

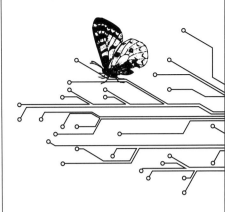

a lot!" she says. She also says that she likes to spend time with her beloved shop cat, MOSFET, but then, she does a lot of that on-air, making the cat part of the programming. Spend any time with Limor and you get the distinct impression that there are few firewalls between her job and her private life. When I asked her what she does in her downtime, besides playing with MOSFET, she offers: "Hacking."

OPEN SOURCE BUSINESS

Besides being a company and community championing open source hardware and software, Adafruit is also a company that has open-sourced many of their business operations to share with the rest of the maker pro/small business communities. Each week, they offer #MakerBusiness posts on the Adafruit blog and chronicle the successes and challenges of an open source maker business in their daily newsletter, AdafruitDaily.com. "We discuss everything, from how many packages we ship, to helping makers consider things like getting a trademark for their company," says Limor. "We consider this a public service that we provide. There were few, if any, such resources when I started, so this is us giving back."

Special Section — Board Guide 2017 | Adafruit

Adafruit TV Guide

Adafruit hosts a number of highly regarded YouTube shows (youtube.com/user/adafruit). Here are a few of our faves:

ELECTRONICS SHOW AND TELL (Wednesdays, 7:30pm ET, G+ Hangout On Air) – A weekly Hangout where Adafruit customers, from young makers to seasoned engineers, show off their projects and what they've built using Adafruit products. Ladyada gets many ideas for new products from these shows.

ASK AN ENGINEER (Wednesdays, 8pm ET) – The longest running live weekly web show about electronics and engineering, *Ask an Engineer* finds Ladyada holding court and clearly and patiently answering viewer questions, demoing products, and talking to guest engineers and makers.

LIVE FROM THE DESK OF LADYADA (Check listings for showtimes) – Look over the shoulder of Ladyada as she livestreams herself at work. *From the Desk* is a great way to learn about electronics, circuit design and troubleshooting, writing code, and more.

3D HANGOUTS (Thursdays, 3pm ET) – Join brothers Noe and Pedro Ruiz for 3D Thursday's *3D Hangouts* show to discuss all things 3D printing and desktop fabrication.

COLLIN'S LAB (Check listings for showtimes) – The quirky, brilliant, and always-entertaining Collin Cunningham teaches viewers the basics of electronics, from what components do, to how to assemble basic circuits, to how to use electronics tools. And he does it all in a natty suit.

JOHN PARK'S WORKSHOP (Check listings for showtimes) – Well-known maker extraordinaire John Edgar Park builds whimsical and creative technology projects, such as illusions, costume props, and robots, while teaching you the skills you need to venture into projects on your own.

STATE OF THE FRUIT

Adafruit prides itself on the supportive culture it creates among employees. Limor says she especially loves seeing employees grow and advance, and she takes great satisfaction in offering excellent employee benefits. Limor also notes that Adafruit has been able to bring significant electronics manufacturing to the U.S., specifically, to New York City. The positive, open source culture that Adafruit has engendered has attracted talented engineers and makers from around the world who want to engage with the company, support its community-driven products and culture, and to work there.

Every week there's an all-hands "State of the Fruit" meeting where the entire group convenes for a weekly assessment of what's going on; to brainstorm ideas, talk about upcoming products and initiatives; and to touch base with the entire team. At the end of these meetings, they have something called Hug Reports.

"Hug Reports are the opposite of 'Bug Reports,'" says Limor. "A Hug Report gives everyone a chance to say thank you to someone. Little things matter, and if we're all celebrating each other, it makes our work and our values better. And that's something our community and our customers notice in how we do things." For the hug reports, employees single out someone who's done something special, admirable, above and beyond the call of duty. These can be as global as Biniam Tekola in the kitting department saying: "Thank you for such a diverse company led by a strong woman" to very specific praise, like Dano Wall in the fabrication department thanking fellow fabber Vance Lewis "... for doing a ton of re-work and fixing a bunch of boards that needed some love."

Limor and an 11-year-old future engineer at the HOPE conference.

ADAFRUIT LEARNING SYSTEM

Since the very inception of Adafruit, educating people of all skills levels and interests about electronics and high-tech making has been a top priority. In 2012, the company launched The Adafruit Learning System, a free online resource for learning about electronics. The well-designed and maintained system currently features over 1,000 user-friendly tutorials on electronics, Raspberry Pi, Arduino, Flora/wearable computing, Internet of Things, NeoPixels, 3D printing, LEDs, and other topics of interest to the DIY/maker community.

GREAT MOMENTS ON ASK AN ENGINEER

Every week for the past seven years, makers, hackers, engineers, and nerds of all stripes tune into Adafruit's live YouTube video show, *Ask an Engineer*. On it, Ladyada answers engineering questions live, shows off new products, and does various electronics demonstrations. One of the hallmarks of *Ask an Engineer*, *Show and Tell*, and all of what Adafruit does, is how approachable Ladyada manages to make such complicated subject matter and how diverse the participants are; all ages and walks of life, men and women, boys and girls, and with widely different skill levels. Ladyada tells one of her favorite moments from it:

"I often have my friend Amanda ['w0z' Wozniak] on the show. She's also an engineer. Once, she shared the story on an episode where a parent had emailed to tell her that their 11-year-old daughter, who watched *Ask an Engineer*, had asked: 'Do boys do engineering, too?'" Ladyada says she was also asked the same thing at last year's HOPE (Hackers on Planet Earth) conference by a little girl. "These girls will never know a world where there aren't women doing engineering," she says.

With such devoted and ambitious champions of open source hardware, software, and empowered maker businesses as Limor Fried and Adafruit, we can only hope for this and so much more.

GARETH BRANWYN is a contributor to *Make:*, Boing Boing, and Wink Books. His latest book is a best-of writing collection and "lazy person's memoir" called *Borg Like Me (& Other Tales of Art, Eros, and Embedded Systems)*.

makezine.com/57

Parts Number: 2005

DESCRIPTION
Online learning resource, marketplace, and maker community for do-it-yourself electronics

TECHNICAL DATA
- Adafruit employs 105 people in their 50,000-sq.-ft. factory in Manhattan
- 100% woman owned, no loans, no venture capital
- Recorded $45 million in revenue in 2016
- Received its millionth order in January 2016
- 14 million website page views and over 2 million uniques a month
- 34 million YouTube views and over 207,000 subscribers
- Social media reach: 119,000+ Twitter followers, 2.1 million followers on G+ (4 million for Ladyada), 77,000 Facebook subscribers, 51,000 Instagram followers
- Limor Fried was featured on the cover of *Wired* (April, 2011) and was named Entrepreneur of the Year by *Entrepreneur* magazine in 2012
- Limor is a founding member of the NYC Industrial Business Advisory Council
- Adafruit is ranked No. 11 among the top 20 U.S. manufacturing companies, No.1 in New York City by *Inc.* magazine, and is listed among *Inc.*'s 5000 "fastest growing private companies"
- In 2016, Limor was named one of the White House's "Champions of Change"

FEATURES
Adafruit's 10 Most Popular Products

1. Adafruit Ultimate GPS Breakout — 66 channel w/10 Hz updates
2. Adafruit Motor/Stepper/Servo Shield for Arduino v2 Kit
3. PowerBoost 1000 Charger — Rechargeable 5V Lipo USB Boost @ 1A
4. Circuit Playground — Integrated microcontroller and sensor board
5. PiTFT Plus 480×320 3.5" TFT+Touchscreen for Raspberry Pi
6. Adafruit 9-DOF Absolute Orientation IMU Fusion Breakout — BNO055
7. Adafruit Feather HUZZAH with ESP8266 WiFi
8. Adafruit Feather 32u4 Basic Proto
9. Adafruit Feather 32u4 Bluefruit LE
10. Adafruit Pro Trinket — 5V 16 MHz

[*Source: Adafruit product stats 2/18/17]

APPLICATIONS
Ladyada's 10 Lessons for Building Open Culture Companies

- You can be a good company and a good business.
- Open source isn't a business or a marketing strategy for us, it's the DNA of our company, it's part of what we do.
- Metrics — if you're not measuring things, you cannot improve them.
- We have a weekly all-company meeting called "State of the Fruit." Be transparent with all parts of your business, early and often.
- Skills can be taught. Good people making good decisions should be the focus and what is celebrated.
- Celebrate others. It's not just about you and your products.
- Traveling takes too much time. Use the power of the internet. Publish frequently, from videos to blog posts.
- Say no to things. It's not about what you can do, it's more about what you will not do.
- Get a good trademark lawyer. If you're open source, you're giving away everything but your name, it's important to protect it.
- You do not need a fancy office or building to do great work. Great work can happen anywhere, even in an apartment.

Special Section | **Board Guide 2017** | Open Source Processors

Written by
Krste Asanovic

An Open Chip

THE JOURNEY TO DEVELOP AN
OPEN SOURCE INSTRUCTION SET

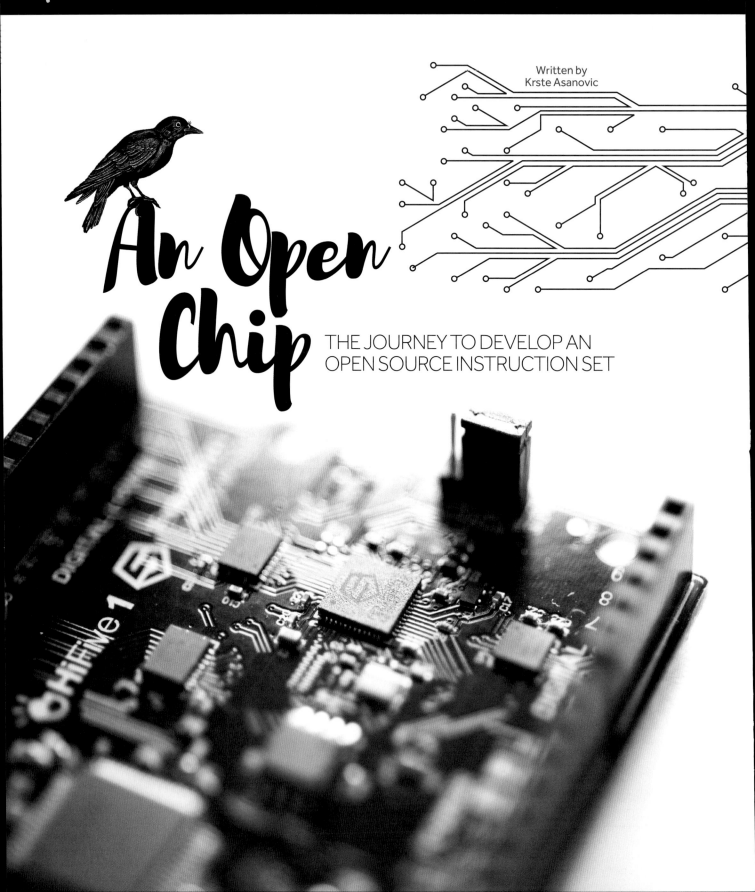

THERE'S BEEN AN UPSWELL OF INTEREST IN CUSTOM, OPEN HARDWARE, in which community-developed and shared designs abound. The availability of low-cost development boards such as Arduino and Raspberry Pi, along with open source software, has driven down the barrier to entry for innovative hardware designs.

But in terms of embracing open source to achieve greater innovation and productivity, the hardware industry is still far behind software. Up until now, the open source hardware movement has been limited by the use of off-the-shelf, commercial silicon chips. These chips often include blocks that are closed source, and whose programming interfaces can only be accessed under nondisclosure agreements or by using opaque precompiled software "binary blobs" that can't be modified or reverse engineered. Many advanced chips are not even available to purchase in small quantities, as the vendors are only interested in supporting high-volume customers. Without access to custom silicon technology, makers and small startups are currently constrained to using off-the-shelf microprocessors combined with field-programmable gate arrays (FPGAs), which can be reprogrammed to emulate a custom chip design. FPGAs, although excellent for prototyping, are too expensive and power hungry to use in large production runs. This lack of open source chips led me, along with my computer architecture research group at University of California, Berkeley, to develop an open source instruction set architecture (ISA). The latest version, RISC-V, allows hardware developers open access and full power over their parts — down to the level of the chip.

ROOM TO IMPROVE

Back in early 2010 our research group was pondering which ISA to choose for our upcoming projects. An ISA defines the set of instructions that a microprocessor understands. For example, laptop and server chips from Intel and AMD only use the Intel x86 ISA, whereas mobile chips from Apple, Samsung, and others only run software encoded in the ARM ISA. Ideally we wanted to run a wide range of software. Using x86 or ARM would have seemed the obvious choice, but three big problems forced us down an alternative path.

First, both these ISAs are large and complex. The Intel x86 has its roots in the 16-bit 8086 chip, quickly designed in 10 weeks in the late 1970s after a different, more ambitious Intel ISA was late to market. In a momentous decision, IBM chose the cheaper Intel 8088 variant of this chip for its first PC prototype and brought in third-party operating software from Microsoft, inadvertently leaving the platform open for clones and enabling Intel and Microsoft to grow into industry titans on the back of the IBM PC platform's dominance. Intel rapidly grew the x86 ISA, expanding from 16- to 32- to 64-bit registers — while preserving backwards compatibility with the original hastily constructed foundation. Nearly 40 years later, not even an Intel architect will claim the x86 ISA is elegant, and high-performance implementations of this baroque ISA require huge engineering resources, far beyond the reach of a small university team.

The ARM ISA, in contrast, has its roots in the RISC (Reduced Instruction Set Computer) movement pioneered by groups at IBM, Stanford University, and Berkeley in the early 1980s, a movement that promoted simple ISAs for high-performance implementations. "RISC" was coined by my colleague Dave Patterson to name the world's first RISC-I and RISC-II chips produced in the project he led at Berkeley, but RISC is now used to refer to the style generally. In 1985, inspired by the student-led Berkeley RISC effort, engineers at Acorn Computer in Cambridge, England, built their own high-performance 32-bit Acorn RISC Machine (ARM) to replace the 8-bit 6502 microprocessor used in their earlier desktops. Unfortunately, the Acorn RISC desktop machines didn't survive the IBM PC onslaught, but Apple picked it for the original Newton handheld project. Apple, VLSI Technology, and Acorn formed a new company, ARM Ltd (ARM now meaning "Advanced RISC Machines"), in 1990 and created a new business model. Instead of selling its own chips, ARM licensed processor designs for customers to incorporate in their own systems on chip (SoCs). ARM now dominates the mobile and embedded SoC processor market and is at the heart of almost every smartphone and tablet. Back in 2010, the ARM ISA didn't meet our needs; it lacked 64-bit addressing and had quirks that complicated high-performance implementations.

The second big problem with both ISAs is that they're not open. The ISA is perhaps the most important interface in computing,

KRSTE ASANOVIC is an EECS professor at University of California, Berkeley, where he leads the RISC-V project. He is also the chairman of the RISC-V Foundation and the chief architect and co-founder of SiFive, a commercial purveyor of RISC-V chips.

To learn more about RISC-V, read the full story at makezine.com/go/risc-v.

as it connects software to hardware, but there have been very few open ISAs, even though many other parts are built around open standards. Apart from Intel, only AMD and VIA can build x86-compatible processors. ARM prohibits other companies to make or sell ARM-compatible processors without a license. We believed that it was important for researchers to share designs, both to improve the quality and reproducibility of research and to share the effort of building research infrastructure. We considered the few existing ISAs but decided against them; they were not open source and lacked 64-bit design, among other technical deficiencies.

The third big problem was the lack of extensibility in these ISAs. Our research project was investigating specialized processors to improve compute capability now that historical Moore's Law scaling was coming to an end. Existing ISAs weren't designed to support much expansion, wastefully using up many instruction bits for the initial ISA design, and after decades of enhancements, there was little encoding space left to add new extensions.

Although we set out to use an existing ISA, we realized that creating our own might be the best path forward, so we began our design in May 2010. We named the new ISA "RISC-V" ("risk-five") to represent the fifth generation of RISC designs from Berkeley.

COMMUNITY CLAMORING

While we worked on the design, we were also working on the supporting software as well as chip implementations. The first RISC-V implementation was the 64-bit Raven-1 chip in 2011, fabricated in a 28-nm FDSOI process donated by ST Microelectronics. Subsequently, over a dozen RISC-V implementations have been completed at Berkeley in a range of fabrication technologies, with more underway. It was also used in a growing number of university classes, with our course material freely posted on class websites.

The RISC-V ISA design was refined over these first few years based on our software and hardware implementation experience and by learning from the mistakes made over the last 30 years of ISA design and use, but we started receiving emails from outside Berkeley complaining about these changes. There was a user community out there relying on RISC-V and our software tools for real engineering projects, and we realized the demand for a free and open ISA standard.

By May 2014, we felt confident enough in the RISC-V base ISA design to freeze it permanently. That summer, in order to publicize it in both academia and industry, Dave Patterson and I wrote a series of position papers on why instruction sets should be free. We could see no technical reason why a free and open ISA standard could not work just as well, if not significantly better, than the proprietary standard. More importantly, we believed a free and open ISA would encourage innovation by allowing for more competition.

For example, an SoC designer currently has to first pick a processor vendor, possibly through a competitive bid, and is then stuck with that vendor's proprietary ISA. Most SoC designers hope their product is successful enough to warrant a second version, but by then, it's generally too expensive to port all the software to another ISA. So the team is effectively locked into the original, proprietary choice, which makes subsequent licensing negotiations much less competitive. With an open ISA like RISC-V, an SoC designer can choose between multiple vendors, use an open source design, or build their own. In fact, when we held the first RISC-V workshop in January 2015, we were surprised to learn that the initial RISC-V processors were already shipping commercially in dental cameras produced by Rumble Development in Massachusetts. Rumble had built its own simple RISC-V softcore to fit in an FPGA in only three weeks.

A SOLID FOUNDATION

The next workshop in June 2015 also sold out with a larger group of attendees. It was clear that RISC-V was now bigger than a university project. In August 2015, I founded the RISC-V Foundation, a 501(c)(6) nonprofit trade association to manage, protect, and promote the RISC-V ISA, and hired Rick O'Connor as the executive director. We launched the foundation in January 2016 with 16 member companies, and by the fourth workshop in July 2016, we'd grown to nearly 50 member companies. Membership ranges from small startups to some of the largest companies in computing, including Google, Microsoft, IBM, HPE, Qualcomm, and AMD. It will ensure that the ISA remains free and open, and uses trademark licensing so that commercial implementations remain compatible with the spec in order to avoid fragmentation. ●

Every Child an Inventor

THE MICRO:BIT FOUNDATION IS SPREADING ITS EDUCATIONAL MISSION AROUND THE WORLD

Written by Zach Shelby

ZACH SHELBY is the CEO of the Micro:bit Foundation, an angel investor and a thought leader in the IoT space.

LAST YEAR, THE BBC COMPLETED AN INCREDIBLE PROJECT that deployed a small, embedded computer to nearly a million school children in the United Kingdom. The goal was to provide the experience of digital creativity and coding to every child, thus increasing interest in STEM careers and confidence with technology. This palm-sized computer, called the BBC micro:bit, integrates modern embedded ARM, Bluetooth, and MEMS directional sensor technology with an easy web interface designed for young children and teachers. (See page 29.)

When I was growing up in Michigan in the 1980s, getting my hands on the Commodore 64 at age 8 made a big impact on my life. Developing my own games — at first, line by line from magazines and then all on my own — was very fulfilling! It led me to learn more about digital circuits and radio, and I eventually became an expert on the Internet of Things and a successful startup entrepreneur. Last summer, I was given the incredible opportunity to create and lead the Micro:bit Foundation, a nonprofit that brings the BBC micro:bit to the entire world. My vision is that, in the future, every child will be an inventor. It's therefore our job as engineers and makers to ensure that the technology is available to help children solve tomorrow's challenges.

We've had an incredible journey building the foundation. Our web page (microbit.org) is available in over 10 languages, and five countries in Europe and Asia have announced national school rollouts. We're also now available in the United States. Creating an entire organization and global infrastructure — and launching publicly in just 3 months — was quite a challenge! My experience with launching and advising startups was helpful. But taking on a project this size called for new solutions, tight teamwork, and hard work alongside 30 partners in order to make sure teachers and students could depend on us for their daily studies.

The thing I love about the micro:bit is the community that has grown around it. Educators, young people, companies, engineers, and makers around the world are volunteering their time to create learning resources and project ideas. They're also contributing to the very micro:bit technology that our own team helps to cultivate and promote. I was recently inspired by Ross Lowe, a 16-year-old micro:bit user who created his own startup making a micro:bit accessory and educational resources for schools.

I can't wait to see what makers will do next with micro:bit!

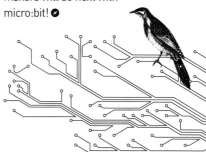

Special Section | **Board Guide 2017** | Voltera Review

Review: Voltera V-One

MAKE CUSTOM PCBS AT HOME WITH LESS MESS

Written and photographed by Matt Stultz

voltera.io • $2,199

MAKING QUALITY PRINTED CIRCUIT BOARDS (PCBS) AT HOME has yielded many solutions through the years with varying degrees of hassle, special equipment, and mess. The most popular DIY systems usually involve the use of acids to etch circuit traces from copper-coated fiberglass boards. The Voltera V-One minimizes the mess and makes DIY PCB production more accurate and automated.

INSTRUCTIONS BUILT IN

The Voltera V-One uses a gantry system, similar to a 3D printer or CNC mill, to move accurately in the X, Y, and Z dimensions. Rather than having a single fixed print head, the V-One has three tool heads that attach magnetically: a probe to measure the blank PCB and feature locations, a conductive ink dispenser that draws the circuit traces and part pads, and a solder paste dispenser that applies solder to pads for surface-mount devices (SMD). The base of the V-One also heats up like a skillet to bake the conductive ink into place and reflow SMD parts.

The Voltera team has done an incredible job of guiding the user through the process of making a PCB on the V-One, with the software shepherding you through each step. After uploading a Gerber file, you mount the blank PCB, measure its location, print the design, and let it bake to set the conductive ink in place. From there, you drill through-holes and vias manually (using more conductive ink to connect the two sides of the PCB). The machine applies the solder paste, but you have to mount the SMD parts by hand. The V-One then runs the board through a proper reflow temperature profile for the supplied solder paste. Once cooled, you have a complete PCB without any chemicals or milling mess.

The V-One isn't without its problems though. It's easy to over-dispense solder, and with larger pads the software can call for too much ink. The ink itself works well with the supplied solder paste for surface-mount soldering, but I found through-hole soldering jobs had a hard time sticking to the conductive ink traces. Also, while the software is easy to follow, don't expect this to be a quick process. Depending on the size of the board, plan on about an hour or two to complete, which is similar to other DIY PCB procedures. There are lots of steps and for an automated tool, lots of manual interaction. I would also love to see a drill attachment to automate the drilling of through-hole points and vias.

PUTTING THE V-ONE TO USE

To really test the V-One thoroughly I needed to do more than just follow their example projects or download a circuit; I wanted to actually make something on my own. In 2009 shortly after founding my first hackerspace, HackPittsburgh, I hosted a Friday night event where we had fun making beeps and boops on an Arduino based synthesizer project called Auduino (code.google.com/archive/p/tinkerit/wikis/Auduino.wiki). Flash forward to today, and some of the members in my current space inspired me to resurrect that project and turn it into a standalone PCB.

I could have simply designed this as a shield to use with an existing Arduino, but I wanted to make it an all-in-one board. I chose to use an Atmega32u4 chip, the same chip found in the Arduino Leonardo and many other Arduino compatible devices. The 32u4 has the ability to natively handle USB communication and can even emulate other device types (like a MIDI device if I wanted to make my synth a MIDI controller).

I designed my PCB using Fritzing, an easy-to-use PCB layout software. Download the Fritzing and Gerber files, bill of materials, and code to make your own here (makezine.com/go/auduino-share).

Most of the steps for this project were the same as the Voltera software itself, so the workflow was basically: print side one, bake, drill holes, fill vias, print side two, bake, apply solder paste, place parts, reflow, place and hand solder through-hole parts.

TIP: If you plan to drill through the PCB into a piece of scrap wood, be aware that the tiny micro-bits used for drilling PCBs can easily snap. I use a spade bit to drill a wide hole in my wood, then center the spot on the PCB that I'm planning to drill through over the hole so I don't have to worry about drilling too deep.

CREATE WITH CONFIDENCE

I really enjoyed making PCBs with the Voltera V-One. I felt like the process of making double-sided boards and Arduino shields (using one of Voltera's pre-drilled blanks) was something I could do with a lot more confidence than with other techniques. If you want to make PCBs on a regular basis, ditch the acid and grab one of these.

And if you build your own Auduino synth, tweet a video of it in action, tagging @make and @mattstultz. We would love to see what you make! ●

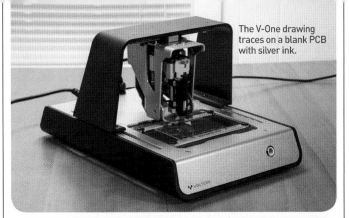
The V-One drawing traces on a blank PCB with silver ink.

MATT STULTZ is the 3D printing and digital fabrication lead for *Make:*. He is also the founder and organizer of 3DPPVD and Ocean State Maker Mill, where he spends his time tinkering in Rhode Island.

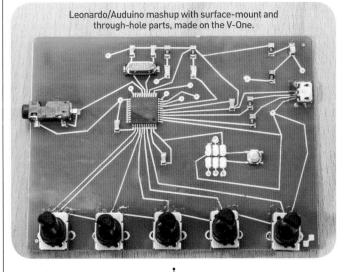
Leonardo/Auduino mashup with surface-mount and through-hole parts, made on the V-One.

Skill Builder

TIPS AND TRICKS TO HELP EXPERTS AND AMATEURS ALIKE

TIM DEAGAN
(@TimDeagan) casts, prints, screens, welds, brazes, bends, screws, glues, nails, and dreams in his Austin, Texas shop. A career troubleshooter, he designs, writes, and debugs code to pay the bills. He's the author of *Make: Fire*, and has written for *Make:*, *Nuts & Volts*, *Lotus Notes Advisor*, and *Database Advisor*.

Written by Tim Deagan

Digital Stencil Design

Use a laser cutter or CNC machine to transform files into spray-paintable templates

48 We Are All Makers

AS A MEANS OF REPRODUCTION, STENCILING DATES BACK THOUSANDS OF YEARS — tens of thousands if you include cave paintings of hands. While it's an ancient practice, stenciling continues to have tremendous modern relevance, especially with new techniques afforded by digital cutting tools now available to stencil artists. Stenciling is an important creative tool for practices as diverse as street art (Figure A) and surface mount soldering.

BRIDGES AND ISLANDS

A stencil is a tool that selectively allows passage of some material to create text or designs. While pigment is the most commonly stenciled material, other products such as solder may also be applied in this way. Stencils allow passage through holes cut out in the material, such as paper, plastic, cloth, metal, etc. More elaborate designs can be created through the use of **bridges** and **islands** (Figure B).

You can cut stencils by hand, but tools to create stencils from digital images have become increasingly common, so let's learn how to use a digital cutting machine (Figure C) such as a Cricut, Silhouette, or CNC router with a drag knife to create a stencil, and then apply that design with spray paint.

It's an important first step to determine how complex of a stencil you want, and if you want to be able to reuse it. Digital cutting machines are commonly used with adhesive backed vinyl. This offers the potential for stencils that have islands without bridges, allowing very complex designs. The trade-off is that these stencils are not reusable. If you want to stencil the image again, you cut another one on your tool.

VECTOR VS. RASTER

In this skill builder we'll make a one-shot stencil without bridges on adhesive vinyl and a multiuse stencil on cardstock. For simplicity's sake, we'll focus on a single color stencil. We'll use Inkscape (inkscape.org), a free tool that runs on Windows, Linux, and Mac OSX, to create our images. Inkscape works with SVG natively. SVG is a **vector image format** and is importable with the Cricut and Silhouette cutting software. Inkscape can also export DXF, which is a more universal format for vector data. Vector images are defined by lines rather than pixels, so they can scale without any loss. Pixel-based formats such as JPG, GIF, BMP, and PNG are **raster formats** (Figure D, left, alongside a vector image, right) and can be converted to vector images in Inkscape with a tracing function.

Stenciling is a black and white operation. In other words, it's like a binary — either the pigment will pass through a section of the stencil or it will not. Not all images translate to this 1-bit mode very well. Line art and text do extremely well, but photographs usually have the most trouble. Converting an image to pure black and white is often referred to as **setting a threshold** (Figure E). Tools like GIMP and Photoshop are excellent for this. Vector tools like Illustrator or Inkscape can set the threshold as part of the tracing process.

ONE-SHOT OR REUSABLE

Regardless of the tools you use to construct your stencil image, you have to determine the answer to the question asked earlier: one-shot or reusable? A one-shot stencil can have islands without bridges (Figure F). You cut the stencil on adhesive vinyl, adhere it to your target surface, and paint over it. When the paint is dry, you remove the vinyl and reveal the stencil. This process can create extremely detailed stencils at the cost of having to cut a new one for each use.

A reusable stencil (Figure G, following page) has to be contained on a single sheet. The islands must have bridges to hold them in place. This reduces the complexity of the stencil and introduces considerations like how well supported small protrusions are and how flimsy the spacing is between items. Generally, you can solve these issues by reducing the detail level of the stencil.

Cutting a reusable stencil requires thought about the material you're cutting. Cardstock is useful, but can tear or get soggy with too much paint. Acetate or other plastic sheets are strong but may cause dripping since they don't absorb any of the paint. I generally prefer using plastic sheets (sold as report covers at the office supply store) and just using a light touch with the spray paint (Figure H, following page).

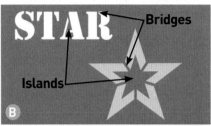

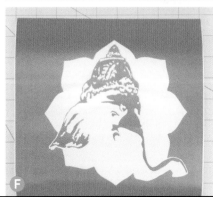

Skill Builder *Digital Stencil Design*

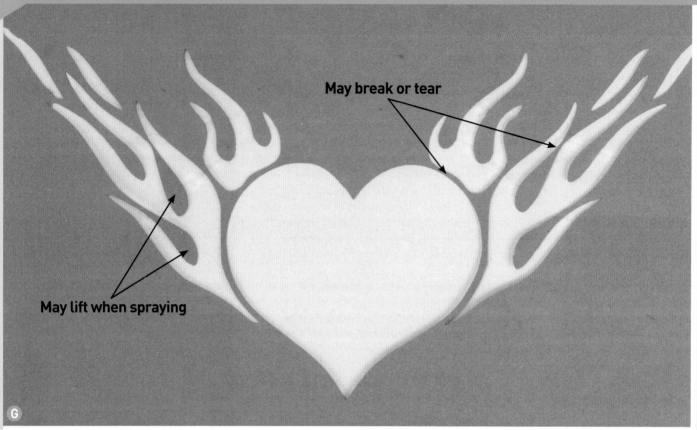

Cutting the stencil on any of the commercially available cutting machines is extremely simple. Load up the program you want to use (Silhouette Studio, Cricut Design Space, and Craft Room are very common) and import your image. It's possible to import vector designs into these tools, but I prefer to import a raster and let the cutting software trace it, because it eliminates a lot of frustration I used to have with hidden and redundant objects in my vector files.

With the image traced, select the appropriate material you're going to cut. Adhesive vinyl usually comes in a roll that you feed into the cutter. Plastic sheet or cardstock is attached to a cutting mat that has a light adhesive which holds it in place. Your software will provide a dialog to select the material and select the presence or lack of a cutting mat. You should also verify that the right cutting tool has been selected. The most important consideration is the thickness, which relates to the depth of cut (Figure I). Ideally you want to cut the vinyl or stock and not the backing sheet or cutting mat.

APPLYING ADHESIVE STENCILS

Adhesive vinyl stencils have a multistep process for application (Figure J):

1. Pick out the parts that need to be removed to let paint pass
2. Adhere transfer paper to the entire stencil (it's like super-wide masking tape)
3. Remove backing sheet and adhere the sticky side of the vinyl to the target. Burnish and remove transfer sheet.
4. Mask around the edges with paper and tape
5. Spray or sponge paint evenly onto the stencil
6. Once dry, remove the stencil. Stand back and admire.

Stencils cut on materials using the cutting mat can just be peeled off for use. If thin parts of the stencil break, or you find that small sections lift off the surface when painting, you can make repairs or enhancements. As you lift away sections of the stencil, move in the same direction as the smaller extrusions, which will help them lay flat instead of pulling them up inadvertently (Figure K).

You can tape breaks, trimming the excess tape away with a utility knife, and you can reinforce parts that lift by taping thread, a wire tie, or a toothpick onto them. You can also assist a stencil's adhesion by cutting holes in areas that you use as a place to apply tape (Figure L).

Stenciling can be done with spray paint, sponged paint, chalk, and other media. Be careful not to let parts of the stencil lift away from your surface when applying your medium, which can cause fuzzy lines. Also be careful when moving the stencil — wet paint under or on the stencil can easily be smeared. If you're using the one-shot vinyl approach, let the paint fully dry before removing the vinyl. Consider putting a coat of clear finish on top of your work to help protect it from scratches.

With stenciling, always remember to be considerate of where you paint your designs, and have fun!

GOING FURTHER

While stencils are generally intended for a single color, they can be created to work in groups, one per color, to create multicolor images (Figure M). Breaking an image into different colored stencils that work together is a task that can be done in image editing programs. Once it comes to the stencil itself, alignment (or **registration**) of the multiple stencils and the order in which they are applied are the only differences between single and multicolor images.

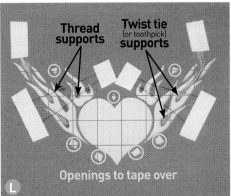

Skill Builder — Programming Robotic Gait

JOSH ELIJAH runs EngiMake, a robotics startup creating educational robotics. His current project is the QuadBot, an open source, animal-inspired robot to teach STEM principles to makers.

Written by Josh Elijah

One Small Step for Robots

Program a Quadruped with Arduino to Get Your Bot Moving

LEGGED ROBOTS ARE GREAT! They can handle terrain better than their wheeled counterparts and move in varied and animalistic ways. However this makes legged robots more complicated, which makes them less accessible to many makers.

Let's get a four-legged robot (also known as a quadruped) walking. I'll take you through a common walk style (called gait) and show you how to program it on an Arduino.

A WORD ON QUADRUPEDS

You'll find quadrupeds abundant in nature, because four legs allow for passive stability, or the ability to stay standing without actively adjusting position. The same is true of robots: A four-legged robot is cheaper and simpler than a robot with more legs, and yet can still achieve stability.

PASSIVE VS. ACTIVE STABILITY

A chair is passively stable, because it doesn't need any control or adjustment to stay upright. A standing human is actively stable because your body requires constant position control to stay standing.

When a quadruped is standing on four legs it is passively stable. When walking, it has options. It can maintain passive stability while walking by keeping three legs on the ground, and reaching out with the fourth. Or it can give up passive stability and use active stability to move faster (albeit less smoothly). These two types of walking gaits are called the creep and the trot. I'm going to show you how the creep gait works.

CREEP GAIT

The creep gait is the easiest walking gait to use. The robot keeps three feet on the ground, and keeps its center of gravity (CoG) inside the triangle formed by those three feet. If the CoG goes outside this triangle for too long, it will fall over (Figure A).

Simple enough. The problem is how to

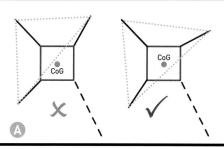

A

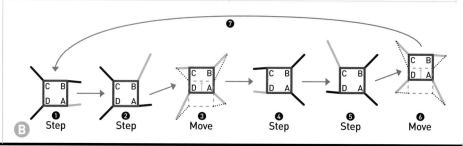

B ① Step ② Step ③ Move ④ Step ⑤ Step ⑥ Move ⑦

52 **We Are All Makers**

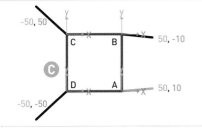
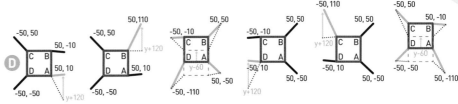
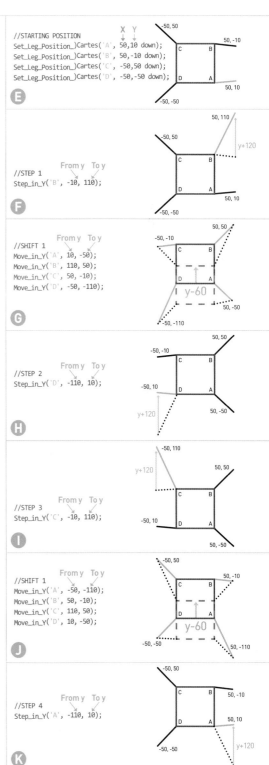

maintain this stability while walking. The pattern in Figure B will save you hours of trial and error (trust me, I know). It's a simple type of passively stable creep gait:

OK LET'S BREAK THAT DOWN

1. This is the starting position, with two legs extended out on one side, and the other two legs pulled inward.
2. The top-right leg lifts up and reaches out, far ahead of the robot.
3. All the legs shift backward, moving the body forward.
4. The back-left leg lifts and steps forward alongside the body. This position is the mirror image of the starting position.
5. The top-left leg lifts and reaches out, far ahead of the robot.
6. Again, all the legs shift backward, moving the body forward.
7. The back-right leg lifts and steps back into the body, bringing us back to the starting position.

Notice that at all times, the triangle formed by the legs on the ground contains the CoG. This is the essence of the creep gait.

When we look at this pattern, we can see it's essentially two sets of mirrored movements. Step, step and shift, followed by another step, step and shift on the other side.

HOW TO WRITE THE CODE

The gait is fairly simple — but how do we actually turn that into code? Well, the first thing to do is decide the specific x,y positions of the legs at each position (Figure C).

Each leg has its own x- and y-axis. We can give the foot of each leg a position, in millimeters, relative to that axis. For example the top-left leg has the position (-50,50). Now we can apply these positions to each stage of the creep gait. Keep in mind that the specific positions you want will depend on the length of your robot's leg. For any arbitrary quadruped you'll need to do some measuring to find the right numbers.

Figure D is an example of the positions.

Between each step, we only need to deal with the change in position denoted by the green arrows on the sequence above. So how does this translate to code? Let's take a look at the Arduino code to implement it, which you can download from makezine. com/go/arduino-quadruped:

```
void loop() {

  //STARTING POSITION
  Set_Leg_Position_0Cartes('A', 50,10,down);
  Set_Leg_Position_0Cartes('B', 50,-10,down);
  Set_Leg_Position_0Cartes('C', -50,50,down);
  Set_Leg_Position_0Cartes('D', -50,-50,down);

  //STEP 1
  Step_in_Y('B', -10,110);

  //SHIFT1
  Move_in_Y('A', 10,-50);
  Move_in_Y('B', 110,50);
  Move_in_Y('C', 50,-10);
  Move_in_Y('D', -50,-110);

  //STEP 2
  Step_in_Y('D', -110,10);

  //STEP 3
  Step_in_Y('C', -10,110);

  //SHIFT1
  Move_in_Y('A', -50,-110);
  Move_in_Y('B', 50,-10);
  Move_in_Y('C', 110,50);
  Move_in_Y('D', 10,-50);

  //STEP 4
  Step_in_Y('A', -110,10);
}
```

Surprisingly simple, perhaps? Let's break it down in Figures E through K.

With this approach you'll have your robot walking in no time. If you use servomotors, you'll need to master inverse kinematics first, which will let you translate servomotor angles into the positions described here. ⏺

PROJECTS
Camera-Mounted Pinoculars

Camera-Mounted "Pinoculars"

Written and photographed by Josh Williams

Snap and view zoomed-in photos through your binoculars, with a Raspberry Pi and Pi Camera

JOSH WILLIAMS helps run makerspaces, loves learning new tools and technology by building and documenting projects, and enjoys exploring the world with his wife.

A

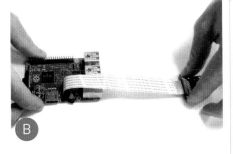

B

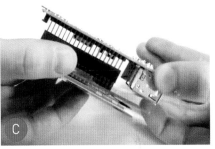

C

Time Required: 30–60 Minutes
Cost: $75–$125

MATERIALS

- **Raspberry Pi 2 single-board computer** with power supply
- **Pi Camera module and cable** NOTE: The default (150mm/6") cable that ships with the Pi Camera is a little tight; you may need to order a longer one.
- **MicroSD card and adapter**
- **PiTFT 2.8" LCD capacitive touchscreen.** Adafruit #1983, adafruit.com
- **USB Wi-Fi dongle or Ethernet cable**
- **Tactile switch, micro** Adafruit #1489
- **Binoculars**
- **Foam, 6mm thick** to mount the camera to the binoculars — wider and longer than the eyepiece of your binoculars
- **Rubber bands, small to medium (2)**
- **Electrical tape**
- **Guide 10 Plus Recharger** Goal Zero #21005, goalzero.com
- **Male USB-A to male USB-micro cable,** ~91cm/36"

TOOLS

- Scissors
- X-Acto or utility knife
- Ruler
- Pencil

Over-Engineered Pinoculars

The version at left involves a laser cutter, computer, Inkscape vector image editing software, thin plywood, nuts and bolts, etc. — and a lot more time. Completely pointless! It is an interesting exercise in designing and fitting a piece of material to an existing object, and the steps are geared toward someone who's never done something like this. Find the project at instructables.com/id/PiNoculars-Raspberry-Pi-Binoculars.

UPGRADE A PAIR OF BINOCULARS (OR A TELESCOPE OR MICROSCOPE) WITH A RASPBERRY PI 2 AND A PI CAMERA and use one of Adafruit's touch LCDs to view and take pictures at a distance.

This project takes more time and money, and is clunkier than if you were to purchase a commercial product — but you're probably not reading *Make:* because you wanted to buy a solution.

We created two versions: an Over-Engineered edition, shown on the facing page and described in the box to the left, and the noob-friendly Quick and Dirty version, detailed here, which involves a little bit of foam, electrical tape, ruler, pencil, and a knife. It can probably be done in half an hour or less. For more information and video instructions, visit this project at instructables.com/id/PiNoculars-Raspberry-Pi-Binoculars.

1. IMAGE DOWNLOAD AND INSTALLATION

The Adafruit crew put together a specific version of the Raspbian OS to make it easy to interface with their 2.8" capacitive touchscreen. Go to learn.adafruit.com/adafruit-2-8-pitft-capacitive-touch/easy-install to download the appropriate one for you.

Unzip the image after it finishes downloading. Carefully follow the "Writing an image to the SD Card" instructions on raspberrypi.org/documentation/installation/installing-images/README.md to install the OS. Eject the SD card once you've finished writing the image to it.

2. TEST YOUR HARDWARE

Remove the microSD card from the adapter. Flip the Pi over, and place the microSD card into its holder. If your camera cable is not attached to the camera, attach it now — blue side facing up, silver tabs facing the lens side (Figure A) — first disengaging the clasp if necessary, and re-engaging once plugged in.

Next, attach the other end of the cable to the Pi, silver tabs facing the HDMI port, blue side facing the network jack, again disengaging the clasp first if needed. Place the cable into the slot and engage the clasp, leaving the camera facing out and the cable hanging over the network jack (Figure B).

Attach the LCD panel to the Pi — be aware that it fits a bit loose and floppy. Line up the GPIO pins on the Pi with the header on the LCD (Figure C). Carefully press down (avoid pressing on the screen itself). Attach power, and boot up. It may take a minute, but you should end up at the Raspi Boot Config — a grey and blue screen.

I used a USB Wi-Fi dongle for this project — it's optional, though the steps later on will make use of it.

PROJECTS | Camera-Mounted Pinoculars

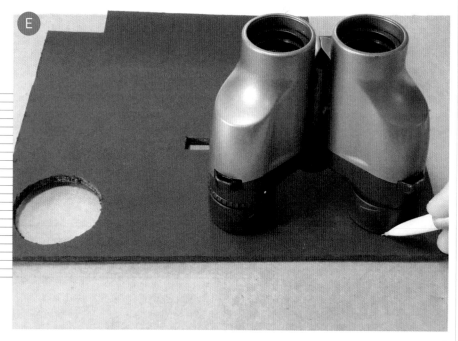

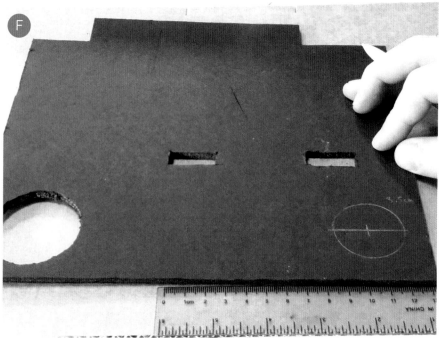

3. OS AND CAMERA SETUP
Boot the Pi. If the Raspberry Pi Software Configuration Tool doesn't run at boot, login and run the command sudo raspi-config to start it.

Go to Expand Filesystem and hit enter. This lets you take full advantage of your SD card. Then go to Internationalization Options and Change Locale (the default is Great Britain), Change Timezone, and Change Keyboard Layout (if necessary) to what is appropriate for you.

Next, Enable Camera, then go to Advanced Options, SSH, and enable the SSH server (Figure D), which allows you to connect remotely. Select Finish and Yes when asked if you would like to reboot.

You have two options to get the Pi connected to your network.
A. Ethernet Cable (Network Cable)
Connect from your router/modem/switch to your Pi. Reboot the Pi. When at the login screen the third line up should be your IP address. Make a note of it.

B. Connect a USB Wi-Fi Dongle to your Raspberry Pi
Run
sudo nano /etc/wpa_supplicant/wpa_supplicant.conf

Add to the bottom of the file:
network={
ssid="YourWiFiName"
psk="YourWiFiPassword"
}\

Reboot the Pi. The third or fourth line up on the screen should be your IP address.

See raspberrypi.org/documentation/configuration/wireless/wireless-cli.md for additional help.

Download and Set Up the Camera Script
Go to learn.adafruit.com/diy-wifi-raspberry-pi-touch-cam/pi-setup and scroll down to "Install Camera." Follow the instructions there to set up Phillip Burgess' excellent camera script.

Then scroll down to "System Tweaks" and follow the steps to speed up the screen redraws so the camera is more responsive. To have the Pi boot straight into the camera software at startup, follow the instructions for "Standalone mode."

Enable Tactile Switch #23
SSH into your Pi: pi@your.pi.ip.address. Then follow the "Tactile switch as power button" steps at learn.adafruit.com/adafruit-pitft-28-inch-resistive-touchscreen-display-raspberry-pi/extras.

NOTE: Do NOT follow the additional steps below the tactile switch section.

4. BUILD THE CAMERA MOUNT
Place your binoculars, with the eyepiece facing down, on the piece of foam. Draw an outline around the eyepiece on the foam with a pencil (Figure **E**). Use a ruler to create a crosshair through the center of your circle (Figure **F**), then draw a rectangle approximately 8mm×8mm at the center (Figure **G**) — this is for the camera lens (Figure **H**). Place some disposable material below the foam to save your worktable, then use your X-Acto knife to remove the rectangle, and then the circle (Figure **I**). Work slowly, making many light cuts; force doesn't always help, and will cause more damage if you slip.

Test-fit the camera mount — it should barely cover the binocular's eyepiece, and the Pi camera should fit just inside the rectangle (Figure **J**). This doesn't have to be perfect, but the closer the camera is to the center of the mount the better.

5. ATTACH THE PI, LCD, AND CAMERA
Place rubber bands around the top and left side of the floppy LCD to hold it in place. Cut a piece of electrical tape long enough (for me, ~25cm) to wrap around the binoculars at least once. Place the sticky side of the tape against the inner edge of the LCD headers. Use your ruler to smooth the sticky side against the headers (Figure **K**).

Making sure the camera cable goes over the network jack, line up and attach the LCD panel to the Pi — being careful not to crush your LCD. Make sure the tape comes out the bottom side (Figure **L**, following page). Then center the Pi and LCD on top of the binoculars, gently pulling the electrical tape

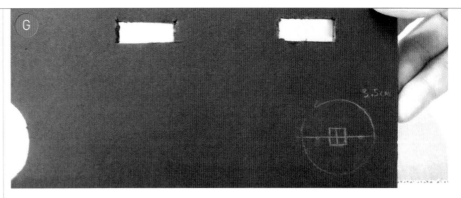

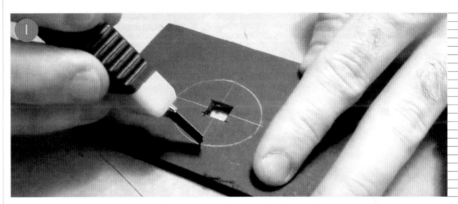

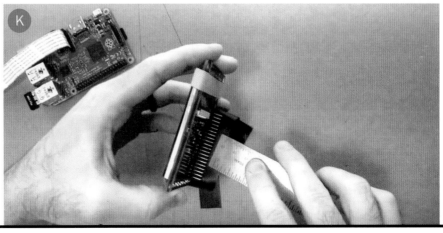

PROJECTS | Camera-Mounted Pinoculars

down and all the way around the binoculars (Figure M). Position it close enough to the front so that the camera can reach the eyepiece. Wrapping loose electrical tape snug around the binoculars worked great with mine. Tape may not adhere well to your binoculars, so experiment as necessary.

Cut a short (~4cm) piece of tape. Place it across the electronics on the back of the camera and wrap it a little bit around the foam mount. If you removed the camera cable for this step, reattach it as you did in Step 2.

Boot up the Raspberry Pi and cut a longer (8–10cm) piece of electrical tape. Pointing the binoculars at a well-lit area, position the camera in front of the left eyepiece, moving it around until you see a well-defined circle (Figure N). Secure the camera to the binocular eyepiece with the tape.

Cut another piece of tape. Wrap that around the eyepiece, over the tape you just attached. Mission complete!

6. CAMERA FLIP!

It's likely that the image you're viewing on the LCD is flipped vertically and horizontally. If this is the case we need to add two lines to the *cam.py* file.

SSH into your Pi (pi@your.ip.address), or plug a keyboard directly into your Pi, and login. Then type:
`cd adafruit-pi-cam-master`
(or type `cd ada` then hit Tab to auto-complete).

Then type:
`nano cam.py -c`
and Enter. Tap Ctrl-W to Search and type:
`# Init Camera`

A 360° View of Your Pinoculars

That should put you around line #571. Not too far above it you should find the line `#Camera.crop` — place your cursor above it and add two new lines:
`camera.vflip = True`
`camera.hflip = True`

Hit Ctrl-X, Y, and Enter to save and exit. Type `sudo reboot` and hit Enter to restart. The video should now be flipped in a way that responds more naturally to your movement.

7. POWER AND MORE

There are a number of options for powering your Pinoculars — we'll cover Goal Zero's Guide 10 Plus Recharger (Figure ⓞ).

Make sure your batteries are fully charged and the battery pack is switched off. Connect the USB cable to both the battery pack and Raspberry Pi. When you're ready to use the Pinoculars, just switch the battery pack on.

When you're done using the Pinoculars, first switch the Pi off using tactile switch #23 on the front of your Pi. Wait at least 30 seconds for your Pi to shut down, then switch the battery pack off.

Some Power Details

The Guide 10 Recharger Pack is rated for 2300mAh at 4.8V. The Raspberry Pi with the LCD and Wi-Fi will probably draw between 500–1000mA per hour, so a 2300mAh battery pack should last around 2 hours. I would recommend against pushing this limit. If the battery can't supply enough power you may end up losing the contents of your SD card, or worse.

If you don't need the Wi-Fi adapter running for this setup, feel free to remove it. It's a pretty significant power drain.

Take close-up photos, like the two above of wildlife, with your Pinoculars.

LEARNINGS AND DIFFICULTIES

Blurry images: Are very common! Holding the Pinoculars as still as possible and shooting in broad daylight will help.

Switch position: Using the touchscreen, or the switches next to it, to take pictures is not ideal. Positioning a separate switch closer to where your fingers rest would be much nicer.

LCD unnecessary? Once you've calibrated the position and focus of the camera lens, the LCD is unnecessary beyond reviewing pictures taken (which is nice). But these could be sent to your phone, or viewed from your phone over a web page.

Video: Taking video with this setup would be nice. I tinkered with adding it briefly, but haven't succeeded. Maybe you will! ⓞ

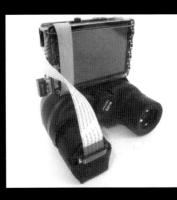

PROJECTS

Open Source Stomp Box

Pedal to the Metal

Written by the ElectroSmash Team

ELECTROSMASH is mad about classic vintage electric guitar effects and loves mixing them up with open source culture, hacking spirit, and new technologies.

Build this open source stomp box and rock out with the best of them

COMPONENTS

Reference	Qty	Value
Capacitors		
C2, C5, C7, C8, C9	5	6.8n
C3, C6, C10	3	4.7μ
C1, C11	2	100n
C4	1	270p
Resistors		
R3, R4, R6, R9, R10, R12, R13	7	4.7K
R5, R7, R8	3	100K
R1, R2	2	1M
R11	1	1.2M
Others		
RV1	1	500K
D1	1	LED 3mm blue
U1	1	TL972 PDIP-8
Socket	1	DIP-8 socket
SW1	1	3DPT footswitch
SW2	1	Toggle switch
SW3, SW4	2	Pushbutton
Conn1,2,3,4	1	40-pin header
J1, J2	2	¼" audio jack

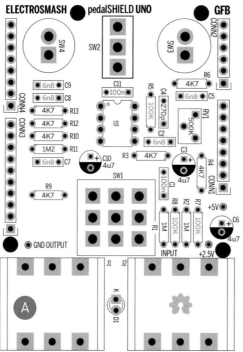

CONSTRUCT A GUITAR PEDAL WITH EASY-TO-FIND PARTS AND HAVE FUN CREATING YOUR OWN SOUNDS IMMEDIATELY — you don't need deep knowledge in digital signal programming or electronics.

This entire project is open source and open hardware; all the schematics and files are free. The design was created using KiCAD, an open source ECAD for Win-Linux-Mac so everybody can contribute. Learn more at www.electrosmash.com/pedalshield-uno. Here's how to build your own.

1. SOLDER THE RESISTORS

There are 13 resistors to be placed, shown in the PCB diagram (Figure Ⓐ) and in Figure Ⓑ:
» 4.7K resistors (7 units): R3, R4, R6, R9, R10, R12, R13.
» 100K resistors (3 units): R5, R7, R8.
» 1M resistors (2 units): R1, R2.
» 1.2M resistor: R11.

In order to solder the components, bend the

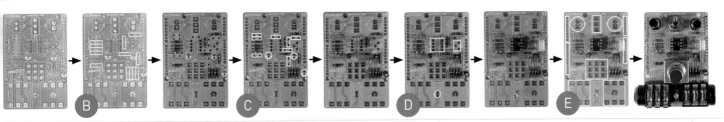

Time Required:
2–3 Hours
Cost:
$80–$100

MATERIALS
- **Arduino Uno microcontroller board**
- **USB cable**
- **Power cable**
- **PedalShield Uno Assembly Kit** includes PCB, all components, and a laser-cut transparent cover, electrosmash.com/store

COMPONENTS IF PURCHASING SEPARATELY:
- **PedalShield Uno PCB** with or without the acrylic cover, electrosmash.com/store
- **Capacitors, ceramic:** 6.8nF (5), 100nF (2), 270pF (1)
- **Capacitors, electrolytic,** 4.7µF (3)
- **Resistors, ¼W:** 4.7kΩ (7), 100kΩ (3), 1MΩ (2), and 1.2MΩ (1)
- **Trimmer resistor (trimpot), 500kΩ** Mouser Electronics #3319W-1-504
- **LED, 3mm blue**
- **Op-amp IC chip, rail-to-rail, TL972 type, PDIP-8 package**
- **DIP-8 socket**
- **Foot switch, 3DPT** aka "stomp switch," Mouser #107-SF17020F-32-21RL
- **Toggle switch, SPDT** Mouser #612-100-A1111
- **Pushbuttons, on-off (2)** Mouser #103-1013-EVX
- **40-pin header**
- **Audio jacks, ¼" stereo (2)** Mouser #NMJ6HCD2
- **Acrylic or other material** for making an enclosure

TOOLS
- **Soldering iron**
- **Wire cutters**
- **Scissors**
- **Pliers**
- **Utility knife**
- **Duct tape**
- **Computer with Arduino IDE software** free from arduino.cc/downloads

leads close to the body, place them through their footprint on the PCB, and solder on the back of the board. Cut the excess leads as short as possible to avoid short circuits. To learn more about how to solder, visit makezine.com/go/learn-to-solder.

2. PLACE THE CAPACITORS
There are 8 film/ceramic and 3 electrolytic caps. Solder them as shown in Figures A and C:
- 6.8nF capacitors (5 units): C2, C5, C7, C8, C9.
- 4.7µF electrolytic capacitors (3 units): C3, C6, C10.
- 100nF capacitors (2 units): C1, C11.
- 270pF capacitor: C4.

Be careful with the electrolytic caps' polarity; the negative lead (the short one) has to be placed in the round hole with the semicircular marking. The positive hole is always square-shaped.

3. ADD THE MEDIUM-SIZE COMPONENTS
Place the DIP socket (U1), 500K trimmer (RV1), and LED (D1) as shown in Figures A and D.

Take care with the LED soldering: The short lead (cathode), on the flat side of the diode, goes into the hole marked with a "K" on the PCB.

4. SOLDER THE BIG COMPONENTS
Place the pushbuttons (SW3, SW4), toggle switch (SW2), 3PDT footswitch (SW1), jack connectors (J1, J2), and header pins (CONN1, 2, 3, and 4), as shown in Figures A and E. Press the op-amp chip into the DIP socket, as shown in Figure F.

When soldering the potentiometers and switches, check their positioning against the plastic cover; it will fit better if it is aligned. Be careful soldering the big components perpendicularly because they tend to be slightly tilted.

Cut the 40-pin strip into 4 segments of 6 pins, 8 pins, 8 pins, and 10 pins. You can use wire cutters, or carve a groove with a utility knife and then just bend it carefully to break it.

Because the USB connector on the Arduino is positioned too close to the output jack, isolate it with some duct tape to avoid short circuits.

5. CHECK YOUR WORK
You should now have a mounted board exactly like

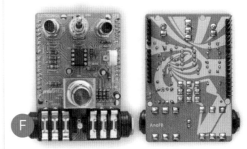

the one shown in Figure F. Double-check your PCB with the model, component by component.

BEFORE POWERING IT UP:
- Visually inspect the bottom of the PCB, ensuring there are no short circuits or long, uncut leads.
- Check that the polarized components (the LED and electrolytic capacitors) are placed correctly.
- Ensure that the op-amp is not upside down.

If you need more help, there's a topic in the forum, www.electrosmash.com/forum/pedalshield-uno, called "Guide to troubleshoot pedalSHIELD UNO."

Finally, the plastic cover can be placed — the footswitch's nut and plastic washer keeps it in place, simple and neat (Figure G). Now just plug the shield into your Arduino Uno. (For performances, consider creating a fully supported enclosure to keep from damaging the electronics.)

NOW ROCK OUT
You can program your own effects in C/C++ or get inspired by our ready-to-play effects: booster, distortion, fuzz, delay, vibrato, tremolo, "Daft Punk" octaver, and more. Grab the Arduino code at www.electrosmash.com/forum/pedalshield-uno/114, plug in, and rock out!

Hear the different sound effects at youtube.com/watch?v=vECPNuPytuw

PROJECTS
Neural Network Cat Spotter

KEITH HAMMOND is projects editor of *Make:* magazine.

SAM BROWN is driven by the need to create, travel, and learn new things. He has a history of making clever wooden things, board games, and circuits, and looks forward to doing stranger things yet.

Cat Activated Laser

Written by Keith Hammond and Sam Brown

This hyperintelligent toy uses furry face recognition to entertain your pet

We Are All Makers

makezine.com/57

Time Required: 3–8 Hours
Cost: $500–$600

MATERIALS

- » **Nvidia Jetson TX1 Developer Kit** Includes the Nvidia Jetson TX1 development board, AC adapter and power cord, USB Micro-B cables (Standard-A and Female Standard-A), rubber feet (4), Quick Start Guide and Safety Booklet, and Wi-Fi antennas (2)
- » **Arduino Uno microcontroller board**
- » **Laser diode, 3V, 5mW** such as Amazon #B00VCR036Q. Or use a 5V laser diode and omit the transistor.
- » **Transistor, NPN**
- » **Hobby servomotors, micro size (2)** such as Amazon #B00ZEDRR3Q
- » **Mounting brackets, for pan-tilt setup** You can fabricate your own from wood, plastic, or metal; buy a pan-tilt bracket; or find one at Thingiverse to 3D print.
- » **USB thumb drives, 32GB (2)**
- » **Mini breadboard** such as Jameco #2155452
- » **Jumper wires**
- » **Enclosures (optional)** for Jetson TX1 and Arduino. We made our own from acrylic; you can download our files and cut and bend them yourself (see HARDWARE Step 2).

TOOLS

- » **Computer** with an x86 processor — i.e., a common desktop or laptop computer, not a Raspberry Pi or other computer with an unusual chip at its heart
- » **Arduino IDE software** free download from arduino.cc/downloads

THIS CAT-TEASING LASER POINTER ACTUALLY RECOGNIZES CATS and only activates when a cat is present — not a person, only a cat!

The cat spotter is built around the Nvidia Jetson, a processor built for neural networks and other AI tasks. It's doing actual visual processing to distinguish cats from all those lesser, non-cat things in the world. Our Jetson loads a neural network that was trained on flashcard-like images until it learned to distinguish felines and other fuzz-balls from reality's more boring, non-furry occupants.

Here's how to build your own neural-network Cat Spotter.

INSTALL THE SOFTWARE

To jump-start you into this project, we've prepared a complete software install on the Jetson, with the Cat Spotter running from the moment it turns on.

The software that updates the Jetson only runs on Ubuntu Linux, so we'll start by making a flash drive that lets any computer boot up as an Ubuntu computer. We used the utility UNetbootin to make this bootable USB stick.

We'll also need the hard drive image that we're going to clone onto a second flash drive — download it from the project page online at makezine.com/projects/jetson-tx1-cat-spotter-laser-teaser.

Both of these downloads, Ubuntu for UNetbootin and the Jetson drive image, are many gigs large, and will take a while to download. Let's construct the Arduino laser controller while we wait.

CONNECT THE HARDWARE

The hardware is simple: a bare laser diode you can buy cheap online, two standard mini servos mounted in pan-tilt brackets, and an Arduino microcontroller to control them. Finally, we add an NPN transistor to let the Arduino switch the 3.3V power on and off.

1. To build a simple pan-tilt laser cat teaser, connect the two servos and the laser diode to the Arduino as shown in the wiring diagram in Figure Ⓐ. The laser is connected to the +3.3V pin; the servos are connected to the 5V pin.

2. Mount the servos to a 90° bracket and

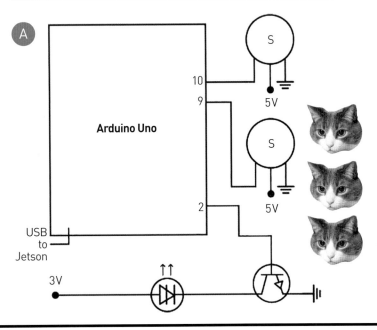

PROJECTS | Neural Network Cat Spotter

to the top of your Arduino enclosure, as shown in Figure B. Then mount the laser to the bracket as well, pointing straight ahead level. After you've gotten the system working, you'll want to tilt the laser to the floor or another surface your cats can reach.

By visiting makezine.com/projects/jetson-tx1-cat-spotter-laser-teaser you can download our Illustrator files to cut the bracket and Arduino enclosure from acrylic (Figure C) or metal and bend them yourself, or download a pan-tilt mount for 3D printing from Thingiverse, or improvise your own solution.

You can also download our Jetson enclosure files for acrylic or metal. (We cut these files from flat acrylic, then used a strip heater to put the bends in them. Note that the PDF files for the enclosure have increased line weights. Many models of laser cutters decide how to cut a line based on its weight, so you may need to change these before cutting.)

3. Download the Arduino sketch from github.com/baudot/cat_play_jetson_triggered. Open it in the Arduino IDE on your computer and then upload it to your Arduino board.

4. Connect the Arduino to the Jetson via USB.

Your laser teaser hardware is complete. Now you just need to program your Jetson to be a Cat Spotter: to recognize cats and initiate the laser teaser whenever it finds a feline.

SET UP THE CAT-SPOTTING SOFTWARE

1. To get the Jetson ready to recognize cats, you'll use a pre-made neural network that's already been trained. We've included this program in the drive image you started downloading before working on the laser.

2. Partition the second USB thumb drive with the ExFAT file system. Most USB sticks use the older FAT file system, which won't accept files over 4GB, such as our 10GB image download.

3. Copy the downloaded file (*laser_cat.tgz*) to the second thumb drive.

4. Reboot your computer with the first

Your New Neural Network

Your new cat toy recognizes felines using a *neural network* — software that learns from experience, similar to the way the brain does. And the Jetson TX1 can run a new image through that network in milliseconds rather than seconds. Superfast image recognition opens up new options for robots that need to keep up with the real world and dodge obstacles.

There's quite a bit going on behind the scenes to make your Cat Spotter run, and all these parts are now installed and ready for you to explore on your Jetson:

Caffe (caffe.berkeleyvision.org) — the neural net we're using to recognize felines is one of the examples included with Caffe, a package for building new neural nets. It's particularly good for visual recognition.

Digits (youtube.com/watch?v=jUiudfxjdr8) — another way to set up and explore neural nets, no coding required.

CUDA (developer.nvidia.com/cuda-education-training) — unlocks the power of the Jetson to run many, many tasks simultaneously, such as …

cuDNN (developer.nvidia.com/cudnn) — the software that taps into CUDA to run neural networks with exceptional speed.

64 We Are All Makers

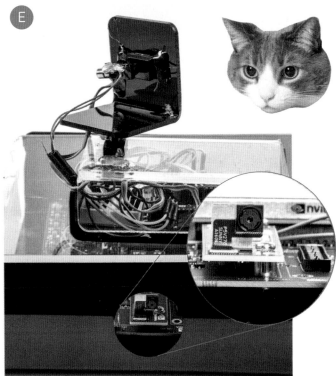

thumb drive, the one prepped with UNetbootin, to launch Ubuntu Linux on your computer. The NVIDIA software tool, JetPack, is written for Ubuntu Linux.

5. Add support for ExFAT drives to that Ubuntu system by opening a terminal window and entering these three commands, in order:
`sudo add-apt-repository universe`
`sudo apt-get update`
`sudo apt-get install exfat-utils exfat-fuse`

6. Plug the second flash drive into the Ubuntu computer, which should now recognize it.

7. Copy the *laser_cat.tgz* file to the Ubuntu system. This copy could take several minutes.

8. Open a terminal, and `cd` to the directory where you dropped the copy of *laser_cat.tgz*. Unzip the file with the command:
`tar -xvzf laser_cat.tgz`
This unzip could take several minutes.

9. After the file has been unzipped and untarred, run the command:
`cd bootloader`
to enter the freshly unzipped directory.

10. Attach the Jetson to your computer running Ubuntu, using the USB to micro-USB adapter cable, with its micro-USB side plugged into the back of the Jetson.

11. Power down your Jetson. Power it back up, immediately holding down the Update button after releasing the power button. While continuing to hold down the Update button, briefly tap the Reset button, wait 2 full seconds, and then release the Update button.

The Power button is the far right button, if the buttons are turned to the front of the Jetson (with the Wi-Fi antenna at the back). The Update button is second from the right. The Reset button is furthest left (Figure D).

12. Confirm that your computer running Ubuntu can see the Jetson attached to it, by running the `lsusb` command in a terminal window. If you see "NVIDIA" in the list that appears, you've confirmed that your computer can connect to the Jetson to update it.

13. You're still in the *bootloader* directory on your Ubuntu machine? Good. Now load our pre-made laser-cat software from the Ubuntu computer onto the Jetson by typing:
`sudo ./tegraflash.py --bl cboot.bin --applet nvtboot_recovery.bin --chip 0x21 --cmd "write APP laser_cat_APP.img"`

Allow this command to run to completion (about 15–20 minutes) and then you're done. You've copied over a complete clone of our working Jetson system, which has the pre-trained neural network.

Disconnect your laptop from the Jetson, and reboot the Jetson. Make sure its camera has a clear view through the enclosure (Figure E). It should immediately start looking for cats!

Getting a Glimpse

For debugging, you can see what your Cat Spotter sees by launching the `imagenet-camera` program from a terminal window on the Jetson. The program lives at *~/Desktop/workspace/cat-spotter/build/aarch64/bin/imagenet-camera*. You'll need to `cd` to its actual directory to run it. When you run the program in user mode, it brings up a video window showing what it sees, and whether it's recognizing a cat at that instant or not. You'll also likely want to tilt your pan-tilt unit so that it always points the laser at the floor, or modify the Arduino code to achieve the same effect.

To see the Cat Spotter in action, visit makezine.com/projects/jetson-tx1-cat-spotter-laser-teaser.

PROJECTS
Mason Jar Fairy Lights

Tricolor
Twinkle Lights

Brighten your July 4th festivities with jars of sequentially fading strands of red, white, and blue LEDs

Written by John M. Wargo

JOHN M. WARGO loves code. He's a professional software developer and author who's into software more than hardware. You can find him on Twitter at @johnwargo.

Time Required:
1 Hour
Cost:
$45–$65

MATERIALS

- » **Adafruit Pro Trinket microcontroller board, 3V 12MHz** Adafruit #2010, adafruit.com
- » **Adafruit Pro Trinket LiIon/LiPoly Backpack Add-On** Adafruit #2124
- » **Lithium ion polymer battery, 3.7V 500mAh** Adafruit #1578
- » **Toggle switch, SPST**
- » **USB cable**
- » **Stranded wire, 22 or 24 gauge** just a few inches
- » **Mason jar**
- » **LED wire strands, 12 LEDs each: red (1), white (1), and blue (1)** Adafruit #897, 894, and 895
- » **Heat-shrink tubing**

TOOLS

- » Soldering iron and solder
- » Utility knife
- » Breadboard
- » Wire cutters
- » Wire stripper
- » Phillips screwdriver
- » Electrical tape
- » Drill (optional)
- » Standoffs (2) and screws (optional)
- » Two-sided tape (optional)
- » Computer with Arduino IDE software free download from arduino.cc/downloads

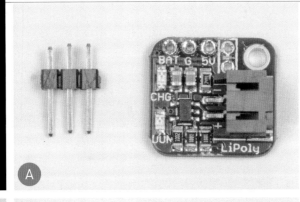

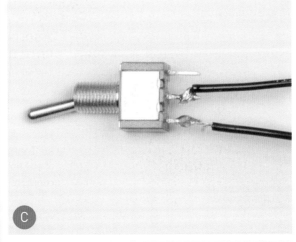

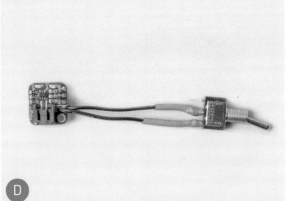

THIS PROJECT IS AN ARDUINO VARIANT OF THE PIMORONI FIREFLY LIGHT PROJECT, made patriotic for the Fourth of July. Basically you place three strings of battery-powered LEDs into a glass jar and use a microcontroller to sequentially fade the light strands — first red, then white, and finally blue, before repeating the series. The fade process repeats for as long as the microcontroller has power.

When I discovered the project, I thought it was really cool, but they threw a lot of hardware (a Raspberry Pi, two HATS, and a battery) into the solution, so I wanted to see if I could build something with fewer parts and for less money. This version of the project uses an Arduino-compatible board (you could use most any Arduino for this project), a battery module, and a battery.

1. BUILD THE LITHIUM BATTERY BACKPACK ADD-ON

To enable the use of a SPST toggle switch to power the board on and off, use the utility knife to cut the trace connecting the two holes with the white box around them on the top right side of the Adafruit Pro Trinket LiIon/LiPoly Backpack Add-On (Figure A). Severing this trace disconnects the board's power output, but once you connect a switch between those two holes, you'll be able to control power to the Trinket board.

Solder the included headers to the Backpack Add-On. The header pins connect to the three holes in the upper-left corner of the board. Insert the headers through the board from the bottom with the black plastic part flush against the bottom of the board. The easiest way to do this is to insert the header pins in a breadboard, then mount the board on top for soldering (Figure B).

2. WIRE AND SOLDER THE TOGGLE SWITCH

Cut two short lengths of wire, then strip the ends and solder one end of each wire to the switch (Figure C).

Finally, solder the open wire ends to the two switch holes — where you cut the connecting trace in Step 1 (Figure D). Since the hardware will be bumped around in the glass jar, I placed heat-shrink tubing around the solder connections to protect them from being shorted out.

PROJECTS | Mason Jar Fairy Lights

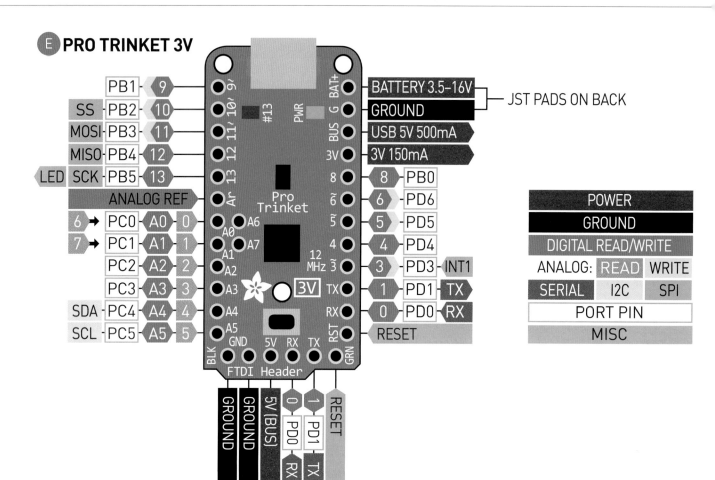

E) PRO TRINKET 3V

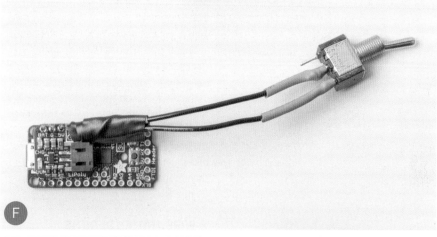

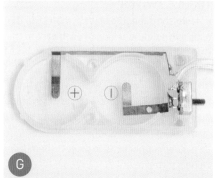

3. SOLDER THE BACKPACK ADD-ON ONTO THE PRO TRINKET

The battery board mounts to the Battery, Ground, and USB 5V pins in the upper-right corner of the Trinket, as shown in Figure E.

Figure F shows the completed board assembly. To protect the USB connector on the Trinket board, I also placed a small piece of electrical tape on top of the board's USB connector; it's visible in Figure H.

4. DISASSEMBLE THE LIGHT STRAND UNITS

Each strand comes with a battery pack and switch, but since we're driving them all from the Trinket, you won't need those parts.

For the Adafruit color LEDs, use a Phillips screwdriver to remove the screws from the battery case and open the case. The LED strand connects to the switch and battery contacts as shown in Figure G. Make note of which wires are connected to the battery's positive (+) and negative (−) terminals. In my case, the contact in the left battery slot connects to the battery's positive (+) terminal and the one in the right connects

Adafruit, Sydney Palmer

to the negative (–) terminal. I could tell this because the case was embossed with little + and – signs.

Next, cut the wires. If they aren't coated in a way to make it easy to tell the difference between the two, quickly tie a knot in the negative wire. ("Knot" = negative. Get it?) That way you'll know which is which.

5. SOLDER THE LED STRANDS TO THE TRINKET BOARD

Solder the wires to the appropriate pins on the Arduino Pro Trinket board. For each of the three strands, solder the negative (–) wire to one of the two GROUND pins (two to one pin and one to the other) in the lower-left corner of the board (see Figure E). Solder each strand's positive (+) wire to pins 9, 10, and 11, one strand to each pin.

Plug the lithium polymer battery into the LiIon/LiPoly Backpack Add-On.

6. DOWNLOAD THE CODE

Connect the Trinket to your PC using a USB cable. Download the project's compiled code from github.com/johnwargo/Arduino-Twinkle-Lights-Array, then upload it to the Trinket. You can find complete directions on how to set up your Arduino development environment and deploy compiled code in the Trinket tutorial (learn.adafruit.com/introducing-pro-trinket/overview).

7. MOUNT IT ALL IN A JAR

You could just stuff the whole contraption into the jar — letting the hardware sit at the bottom. To charge the battery, or flip the switch, you'd have to pull the whole assembly out of the jar and put it back when you're done.

However, the Adafruit lights are rather small and the strand is short, so the hardware stands out more prominently. The Mason jar implementation provides an interesting option for mounting the hardware. I drilled some holes in the lid, and using two standoffs and screws, mounted the hardware to the underside of the lid. I then used two-sided tape to adhere the battery to the lid, underneath the microcontroller (see Figure H). With this approach, you simply remove the lid to charge the battery or flip the switch, reducing stress on the assembly.

To operate, just turn it on and close the lid. You're done! ◉

The Project's Code

Instead of defining a *variable* for each output pin, this Twinkle Lights application uses an *array* to hold the output pin values:

```
//Analog output pin assignments
//Populate this array with the 1st of pins you've connected
//LED strands to.
int ledPins[] = { 9, 10, 11 };
//Update the numPins constant to match the number of
//elements in the array
const int numPins = 3;
```

Populate the `ledPins` array with the list of analog output pins your LED strands are connected to (the strand's positive wires). These pins don't have to be in sequence like the example shown here. Connect the strands to whichever analog pins you want, just make sure to populate the array in the order in which you want the pins used. Next, update the `numPins` constant with the number of strands (pins) you're using.

I added a function to get the next array index:

```
int getNextIndex(int currentIndex) {
  //Get the next array index
  //start by incrementing the current pin
  int idx = ++currentIndex;
  //does pin exceed the bounds of the array?
  //I could use a == comparison here as pin should never go above
  //numPins, unless I have a logic error in the code, but I'm going
  //to check anyway
  if (idx >= numPins) {
    //Then reset to the beginning of the array
    idx = 0;
  }
  return idx;
}
```

Before starting each loop iteration, the `upPin` and `downPin` variables are updated to point to the next elements in the `ledPins` array. This enables the application to maintain a *sliding window* into the array — pointing to two sequential analog output pins as it iterates through the array. The loop is greatly simplified as well:

```
void loop() {
  //Increment our pin designators
  downPin = getNextIndex(downPin);
  upPin = getNextIndex(upPin);

  //Loop through the voltage output values (ranging from 0 to 255)
  //incrementing by 1
  for (int i = 0; i <= maxAnalog; i++) {
    //Drive the upPin up to maxAnalog
    analogWrite(ledPins[upPin], i);
    //While simultaneously driving downPin down to 0
    analogWrite(ledPins[downPin], maxAnalog - i);
    //Pause for a little while
    delay(delayVal);
  }
  //Wait a second before continuing
  delay(1000);
}
```

PROJECTS | Remaking History

The Birth of Medical Hygiene

Ignaz Semmelweis and his maternity ward revelation

Written by William Gurstelle ■ Illustration by Peter Strain

CAUTION

» While working with sodium hydroxide, always wear safety glasses as well as rubber gloves. Work in a room with good ventilation. Lye can cause chemical burns and may be fatal if swallowed.

» When you mix sodium hydroxide with water, it produces considerable heat. Always use a heat-safe mixing container that is only partially filled. It's best to mix ingredients over a sink.

» When you mix sodium hydroxide and water, always add lye to water. Adding water to lye can cause the mixture to suddenly erupt or boil.

» Don't use aluminum pans or mixing containers because they will react with lye, discoloring the aluminum and ruining the soap.

WILLIAM GURSTELLE's new book series *Remaking History*, based on this magazine column, is available in the Maker Shed, makershed.com.

Time Required:
2 Hours to make the soap,
10 Days til it's ready to use
Cost:
$5–$15

MATERIALS
NOTE: Soap making requires precise measurement. A bit too much or too little of any ingredient can ruin your soap. Measure ingredient weights carefully, using a digital scale.
» **Lard, 16oz**
» **Lye, 2oz** available at most hardware stores as drain cleaner
» **Water, 5¼oz**

TOOLS
» Digital scale
» Pyrex measuring cup, 16oz
» Pot, 3-quart, non-aluminum
» Eggbeater, non-aluminum
» Spoon, wooden or plastic
» Glass baking pan, 8"×8"
» Damp cotton cloth
» Safety glasses
» Rubber gloves

HAND WASHING IS AN IMPORTANT PART OF EVERY MODERN PHYSICIAN'S PRE-PATIENT CONTACT ROUTINE. It's also something that food preparers do when they enter the kitchen, and something that you hope the person that you just shook hands with did when he or she last left the public bathroom. In fact, hand washing as a means to prevent passing on infection is such an ingrained routine that it's hard to imagine a time when it wasn't a common part of daily life.

HOW FAR WE'VE COME!
Only 150 years ago, every location that we now think of as being clean and sanitary, from food preparation areas to hospital operating rooms, was absolutely loaded with bacteria. Doctors would walk from one bloody, suppurating patient to another with little more than a wipe of a dirty handkerchief. It was a filthy, germy world until Ignaz Semmelweis, and public health scientists such as Joseph Lister and Louis Pasteur who followed, cleaned up our public health act.

In 1847, Dr. Ignaz Semmelweis was the chief physician of a two-ward maternity clinic in Vienna. When he reviewed the statistics regarding the condition of mothers in his hospital, he was unable to account for the fact that one ward had a 500% higher rate of a common and deadly condition called puerperal fever.

Such a huge variance in infection rates could not, he reckoned, be due to chance alone. He noted the ward with the higher mortality rate was staffed by doctors while the other was staffed by midwives. Semmelweis spent months analyzing the situation, but was unable to figure out why male doctors were so much more deadly to mothers than female midwives.

Finally, the light-bulb moment came. There was indeed a big difference between the ward staffed by doctors and the midwife-staffed ward: the doctors performed autopsies and the midwives did not. Semmelweis conjectured that small pieces of matter (this was before Louis Pasteur developed the idea of germ theory) were being transferred from the corpses in the autopsy room to the mothers in the childbirth room and this uncleanliness was the cause of infections. Semmelweis' solution was straightforward: physicians were to wash their hands thoroughly with soap between patients!

When Semmelweis' directive was enacted, infections decreased dramatically. Unfortunately, the medical community was slow to adopt his hand-washing advice and years went by before it became mainstream practice.

Even today, people catch colds, flu, and other diseases that could have been prevented by more frequent hand washing. In this edition of Remaking History, we pay tribute to Dr. Semmelweis by making our own batch of disease-fighting soap.

DO-IT-YOURSELF SOAP MAKING

First, be sure you understand that sodium hydroxide, or lye, is a powerful chemical and must never contact skin, eyes, lips, wooden table tops, and so on. It is a strong alkali that reacts with the molecules in fat to turn it into soap. The technical term for this chemical process is *saponification*. Many people believe that "lye soap" is harsh and tough on skin, but if you measure your ingredients carefully, all of the alkali should be used up in the saponification reaction and the end result should be a gentle but effective soap.

1. Don your safety glasses and rubber gloves. Measure out the required quantities of water, lye,

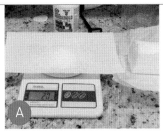

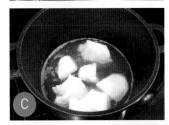

and lard using the digital scale (Figure A).

2. Carefully dissolve the lye into the water in the Pyrex measuring cup. Stir with a spoon until lye is fully dissolved (Figure B). Remember, the lye and water mixture will heat up by itself. Allow the mixture to cool to room temperature.

3. In a large, non-aluminum pot, melt the lard on a stovetop (Figure C) at very low heat until it liquefies. Let the liquefied lard cool until it reaches a temperature of 95°F. Now, add the lye-water solution to the fat, stirring as you pour (Figure D).

4. Saponification, the chemical process in which fat and alkali change to soap molecules, occurs only when the ingredients are very thoroughly mixed. Using a wooden spoon or eggbeater, mix the lye, water, and lard for 15 to 30 minutes (Figure E). Eventually, the mixture will achieve the consistency of honey. When the lard-lye mixture dripping off the stirring spoon or eggbeater leaves a line that remains visible on the top of the mixture for a few seconds (Figure F), the soap is ready for pouring.

5. Line the bottom of the 8"×8" glass pan with a damp cotton cloth. Pour the mixture into the cloth-lined pan (Figure G) and cover the top of the pan with a few sheets of newspaper. Allow the mixture to rest undisturbed for 24 hours.

6. Cut the soap into bars (Figure H). Then complete the saponification process by allowing the soap bars to rest uncovered for 10 days before using.

USING YOUR SOAP

Got soap? Great! Here's some advice from the Centers for Disease Control about how to wash your hands with your Semmelweis-inspired soap:
» Wet your hands with clean, running water (warm or cold), turn off the tap, and apply soap.
» Lather your hands by rubbing them together with the soap. Be sure to lather the backs of your hands, between your fingers, and under your nails.
» Scrub your hands for at least 20 seconds. Need a timer? Hum the "Happy Birthday" song from beginning to end twice.
» Rinse your hands well under clean, running water.
» Dry your hands using a clean towel, or air-dry them.

PROJECTS | DIY Journal

Traveler's Notebook

Craft your own cover and paper inserts for a customizable Midori-style journal

Written by Linn from Darbin Orvar

LINN FROM THE DARBIN ORVAR YouTube channel (youtube.com/user/darbinorvar) is originally from Sweden and has been making things all her life. She enjoys working in her shop on a variety of projects from electronics to woodworking, using hand tools and power tools.

Time Required:
1–2 Hours
Cost:
$20–$40

MATERIALS
FOR THE LEATHER COVER:

» **Leather, 3–4oz medium weight, 10½" × 8½" (266×215mm) account for extra for trimming.** You can often find great deals on scrap pieces in leather stores, or purchase a small piece of leather or faux leather at a craft store.
» **Elastic cord, 50" (1.2m)**

FOR THE PAPER INSERTS:
» **Cardstock paper (3)** 1 sheet for each book
» **Paper (36)** 12 sheets for each book. If desired, use or print your own graph paper, lined paper, or keep it natural.

TOOLS
FOR THE LEATHER COVER:
» **Cutting mat**
» **Square ruler**
» **Tool** to cut the leather such as a rotary cutter, utility knife, or scissors
» **Ruler**
» **Leather hole punch, small**
» **Mallet**
» **Fabric marking pencil or white chalk**

FOR THE PAPER BOOKS:
» **Cutting mat**
» **Paper cutter**
» **Paper clips, large** or small clamps
» **Mallet**
» **Awl**
» **Thread**
» **Needle**
» **Utility knife**
» **Scissors**
» **Ruler**

THE APPEAL OF JAPANESE STATIONER MIDORI'S TRAVELER'S NOTEBOOK LIES IN ITS SIMPLICITY: a beautiful leather cover and simple elastic strap system enable you to keep several notebooks with you, and replace as needed.

Making your own Midori-style traveler's notebook is great because you have your choice of material and size, plus it's a wonderful project for beginners as well as more experienced makers. You can either create the leather cover and purchase booklet inserts, or make the booklets yourself, of your own design.

MAKE THE LEATHER COVER
1. CUT THE LEATHER
Mark out a rectangle measuring 10½"×8½" (or 266×215mm) and cut to size using a rotary cutter, scissors, or utility knife (Figure A).

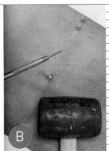

2. MAKE THE HOLES
Measuring out perpendicular from one of the long sides, mark spaces for the 5 holes along the center of the leather at ½", 1", 4¼", 7½", and 8". (13mm, 26mm, 108mm, 190mm, and 203mm). Use a punch and mallet to create the holes (Figure B).

3. ATTACH THE CORD
Cut 30" (760mm) of elastic cord. Fold in half and insert in the middle hole from the inside of the cover, leaving a loop on the outside. Pull each end through the hole closest to it and back in again through the farthest holes (Figure C). Make sure you still have some folded cord sticking out from the middle — this will be the loop you slip over the cover to hold it closed. Tie the two ends of the cord together on the inside of the cover.

Cut another 17" (432 mm) of elastic cord, tie together in a loop, and set aside.

MAKE THE BOOK INSERTS
1. FOLD AND TRIM THE PAPER
Fold the cardstock cover and the individual sheets in half. Using a paper cutter, cut the cardstock and each piece of folded paper to size: 8"×4½" (or 203mm×114mm).

2. ASSEMBLE
Place 12 sheets of paper together to form a booklet, and make 3 booklets. Cover each booklet with the cardstock sheet.

Unfold and lay the booklet flat, then pinch the sheets together using 4 large paperclips or small clamps.

3. SEW TOGETHER
Mark out holes along the fold line of each book (Figure D). Start ½" (13mm) from the top, with 1" (25mm) between each mark, for a total of 8 marks. Using the awl and the mallet, make holes all the way through the folded booklet.

Thread a needle and sew the book together. Starting from the middle inside the book, stitch out towards one end. Then come back (Figure E), overlapping the stitches, and tie the thread together and cut.

4. CLEAN THE EDGES
Place a ruler near the edge of the notebook, hold down firmly, and use a utility knife to carefully trim the protruding sides of the paper (Figure F). Use a fresh blade, and don't apply too much pressure. Repeat several times until you have a clean edge.

ASSEMBLE THE NOTEBOOK
Choose the order in which you want the booklets to appear in the finished notebook, then place your first and third booklets back-to-back and slip on the elastic strap you set aside earlier so that it holds the booklets together spine-to-spine. Slip the second booklet into the middle of the leather cover so that the elastic cord sits in the middle of the booklet. Slip the connected first and third booklets behind the second booklet (Figure G). Close the cover, and fold over the elastic strap. If the strap is too loose, tie it a little tighter.

USE IT!
Now your Midori-style traveler's notebook is ready to use. Write, draw, sketch, plan! Develop a system for how you work, and what you'd like to organize. You can take this concept further and create calendars, dividers, and more, and you can sew a fabric insert to slip underneath the books with pockets to store money, receipts, and notes. Best of all — when you use up a notebook, replace it with a fresh one, and keep using your leather cover as it wears and becomes more beautiful over time.

For video instructions, visit makezine.com/go/diy-midori-notebook

PROJECTS | Air Hockey Robot

Air Hockey Robot

Build your own open source, 3D-printed opponent

Written by Juan Pedro and Jose Julio of JJRobots

NEVER BE WITHOUT A WORTHY ADVERSARY WITH THE AIR HOCKEY ROBOT EVO. It's easy to set up and deploy, and it's controlled by your Android smartphone.

This open source robot can be scratch-built using readily available components and 3D-printed parts, or you can buy it as a kit. It requires an air hockey table — we have one we recommend that fits our bot, but you can modify the dimensions to accommodate others.

It's a fun project that teaches science, robotics, computing, visual recognition, and engineering.

HOW IT WORKS

The smartphone is the robot's brain. Running the Air Hockey Robot app, it processes the data captured by its camera in real time, detecting features on the playing court and making attack/defend decisions according to the detected objects' locations and their trajectories (Figure Ⓐ). Your own device can defeat you in a real game!

Uniform illumination is extremely important for the vision system to work, so avoid shadows, reflections, and if you can, fluorescent lighting.

74   We Are All Makers

MODIFIABLE AND PORTABLE

It's easy to adjust the skill level of the robot, to play with children, for example — just go to the app configuration menu and set the difficulty. You can also set it to "manual" mode and control the robot using your finger, and there is a PlayStation 3 Camera version for the robotics enthusiast, which runs on a regular PC and allows you to modify the vision system. The robot is easily removed from the table for transport, or when you'd rather play against another human.

HOW IT'S MADE

The robot is constructed from metal bars, a timing belt, and 3D-printed parts. The electronics consist of the JJRobots Brain Shield, an Arduino Leonardo, 2 stepper motor drivers, and a 12V fan to cool them (Figure B). The robot is locally controlled by the Brain Shield, which dictates its speed and acceleration, sending the appropriate pulses to the stepper motors. For complete instructions visit jjrobots.com/air-hockey-robot-evo.

The key to a flawless working robot is the H-bot system (Figure C); this structure allows the robot to move to any location on its playing field using only two motors (Figure D). The H-bot needs to run smooth and be strong at the same time. The better you set it up, the greater the accelerations you can reach. Here's how the robot mechanism goes together.

ASSEMBLE THE H-BOT

1. ATTACH MOTOR AND ASSEMBLE PULLEYS

Affix the motor to the motor supports with M3×6mm bolts, 3 per motor (Figure E). Then mount 2 of the 623 ball bearings inside each of the 6 pulleys (Figure F) with an M3×25mm bolt. Secure the fan to its support using 4 of the M3×15mm bolts and M3 nuts.

2. ASSEMBLE THE X-AXIS

Take the 2 aluminum pipes and insert both into a lateral slider cap. This is the most important part of this assembly. It can seem quite hard to insert the pipes into the 3D-printed parts, but the sturdiness of the structure will depend on how tight everything is. Twisting the pipes as you're pushing them helps. If necessary, use a hammer (gently) to prod the pipes into the

JUAN PEDRO and **JOSE JULIO** are electronics hobbyists who really want to share ideas and knowledge with everyone who loves DIY robotics.

Time Required: 30–60 Minutes using the kit
Cost: $200–$300

MATERIALS

» **Air hockey table** We recommend the Buffalo Explorer Mini Air Hockey Table from athleteshop.com. A discount is offered through the JJRobots site, jjrobots.com.
» **Android smartphone, with Lollipop OS or higher**
» **Air Hockey Robot EVO kit** from JJRobots
ROBOT KIT CONTAINS:
» **Bolts: M3×6mm (12), M3×10mm (4), M3×15mm (6), M3×25mm (7)**
» **Nuts, M3 (20)**
» **Self-locking nuts, M3 (2)**
» **Wood screws, M2.5×20mm (12)**
» **EVA foam, 2 different colors**
» **Fan, 12V, 4cm**
» **Stepper motors, NEMA17, with cables (70+70cm) (2)**
» **Brain Shield** from JJRobots, or you can create your own using the info provided at github.com/jjrobots
» **Arduino Leonardo microcontroller board**
» **Stepper motor drivers, A4988, with heat-sinks (2)**
» **Power supply, 12V 2A, 5.5mm/2.5mm jack**
» **Linear ball bearings, LM88UU (2)**
» **Ball bearings, 623 (12)**
» **Stainless steel rods, 8mm dia. × 455mm long (2)**
» **Anodized aluminum tubes, 8mm dia. × 43.5mm long (2)**
» **Aluminum square pipe, 12mm×12mm, 71cm long**
» **GT2 timing belt, 300cm long**
» **Double-sided sticky foam tape**
» **Zip ties, 150mm×3mm (5)**
» **3D printed parts (optional)** Add them to the kit for an extra fee, or print them yourself.

TOOLS

» **Computer with Arduino IDE software** free download from arduino.cc/downloads
» **Phillips screwdriver**
» **Scissors**
» **Hammer (optional)**

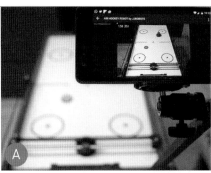

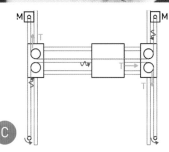

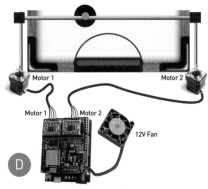

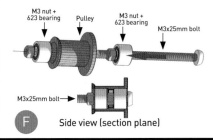

PROJECTS | Air Hockey Robot

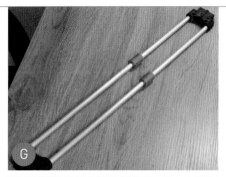
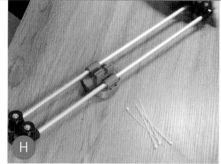

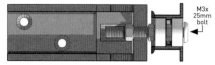

Side view (section plane)

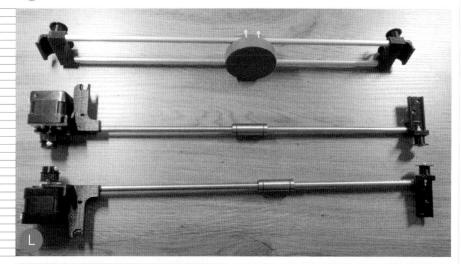

CAUTION: Always, always keep the fan blowing while the Air Hockey Robot is working. The stepper motor drivers might get damaged if there is not a constant airflow cooling them down.

CAUTION: Sudden accelerations of the robot within its playing area could catch you unaware. Before adjusting anything remember to disconnect its power supply.

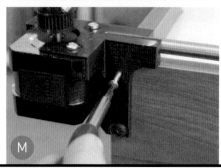
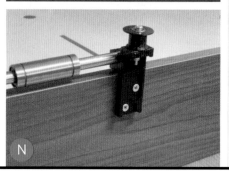

channel. Insert the bushings, shown in red in Figure G, then place the second lateral slider cap on the other end.

Screw on four of the assembled pulleys — no need to use a nut (Figure H). Attach the paddle (aka the goalie, striker, or mallet) in place with 4 zip ties, running them through the two channels on either side of the paddle, and tightening them so as to just hug the plastic bushings. Trim the excess zip ties (Figure I). Test that the paddle slides easily along the pipes. If it doesn't, olive oil will greatly reduce the friction between the aluminum and PLA plastic. Attach the base of the paddle using four M3×15mm bolts. This is the robot's X-axis.

3. BUILD THE SIDE RAILS
Attach a motor pulley to each motor's axis with four M3×10mm bolts and nuts, tightly — these pulleys will transmit all the movement to the robot (Figure J). Affix each of the remaining two pulleys to a side support using a self-locking nut (Figure K). Insert one end of the steel bars into each. Slide the LM88UU linear bearings onto the bars, then insert each into a motor support (Figure L). This completes the H-bot build.

ATTACH TO THE TABLE
At one end of the table, loosen the top screw on the side. Set the motor support in place on top of it, then screw it back in. Use a wood screw to fix it in place (Figure M). Repeat on the opposite side.

Attach the side supports on either side of the table using two wood screws (Figure N). Gently snap the robot's X-axis structure with the sliding pusher onto the LM88UU bearings on the steel bars. The X-axis structure should easily slide up and down the side steel bars. If you find any burr inside the channel where the linear bearing is, remove it gently. Both LM88UUs have to be perfectly aligned in order to avoid any kind of friction with the steel.

For the complete instructions, visit jjrobots.com/air-hockey-robot-evo.

GOING FURTHER
Mount two robots per table, one on each side, and have them compete, or compare different gaming strategies in a tournament. Remember, you can easily change their behavior via the Arduino code. Game on!

Satisfying Spinners

From duct tape to bronze, handmade bearing-based fidget toys are all the rage

Written by Caleb Kraft

Milled Bronze Spinner

WOODEN SPINNER
youtube.com/watch?v=1M5_eMcuSn4
Using wood as a base material can vastly improve the looks.

MASSIVE MDF SPINNER
youtube.com/watch?v=a4zllG7Qjdk
Why only experiment with different materials? This roughly 2' wide spinner provides a much bigger fidgeting experience.

MILLED BRONZE SPINNER
youtube.com/watch?v=AWz8snbgZvM
If you have full access to a machine shop, you can get some impressive results. Milling brass, aluminum, and steel can create professional-quality trinkets.

YOU'VE PROBABLY BEEN ANNOYED BY SOMEONE CLICKING A PEN incessantly as they sit, but many of us can't help but fiddle with whatever we have on hand. This tendency has inspired a wealth of fidget toys.

Fidgeting toys come in all types — we've seen little bumps, knobs, bits that twist together, parts that lock and unlock. The latest craze though, is the "spinner." While the bodies vary, they all have a bearing in the middle. Pinch to hold, give a flick, and enjoy the resultant spin. These little devices feel good, often look good, and best of all, don't make noise to annoy others.

As with any rapidly growing fad, people have taken the construction of fidget spinners from simple to ridiculous. Here are some of our faves.

DUCT TAPE SPINNER
instructables.com/id/Duct-Tape-Spinner-Fidget-Toy
You can make a simple spinner with only a skate bearing, 3 marbles, and a bit of duct tape.

LED SPINNER
instructables.com/id/LED-FIDGET-SPINNER
Some people get creative with their spinners. This design stays fairly simple, but adds LEDs for some cool effects while spinning.

CALEB KRAFT is a wily and fidgety person by nature. This new trend of fidget spinners has let him know that he's not alone and the rest of the planet is possibly as chronically restless as he is. He is also senior editor for *Make:*.

PROJECTS | Fun Zone

Poseable Papercraft
Print, fold, and glue your own Makey mascot
Written by Caleb Kraft

PAPERCRAFT IS A FANTASTIC WAY TO EXPLORE model making. Generally speaking, paper models are cheap and relatively easy to assemble. You may need a steady hand, and a decent amount of time to devote, but the results can be quite stunning.

One person particularly astute in this area is Rob Ives. Once a classroom teacher, Ives has been making paper models professionally since around 2000. He's authored two books: *Paper Locksmith* and *Paper Automata*, both of which incorporate models that function mechanically as well as look good.

We asked Ives if he'd be up to the task of making us a poseable version of our Makey robot mascot and he jumped on the opportunity. You see, every new

Time Required:
3–5 Hours
Cost:
$5

MATERIALS:
» **Paper** Thick paper or cardstock is best, although regular paper will do

TOOLS:
» **Computer and printer**
» **Scissors or utility knife**
» **Scoring tool**
» **Glue** PVA, Tacky Glue, glue stick, etc. depending on thickness of paper

CALEB KRAFT celebrates the act of making in any medium, from steel to ground up trees pressed into thin sheets — aka paper. He's also senior editor for *Make:*.

model and every new job is a chance to obtain a new skill.

In this case, Ives saw an opportunity to learn the common papercrafting software Pepakura. Previously he had done all of his work in Illustrator and by hand, with an expert's eye. He figured it was about time to give this software, which unwraps 3D models for paper re-creation, a try.

He started by modeling the basic shape in Blender (Figure A), an open source and free 3D modeling software. Then he brought this model into Pepakura (Figure B) and began playing with settings. Once he was happy with the template that was dynamically generated, he was able to bring it into illustrator for coloring and further refinement (Figure C).

TO ASSEMBLE

1. Download and Print the PDF
Find the PDF at makezine.com/go/papercraft-makey — there will be 4 full-sized sheets (Figure D). To make him more sturdy and rigid, it is recommended that you print Makey on thick paper or cardstock. If you don't have any available, regular printer paper will do.

2. Cut Out the Pieces
Trim along the solid lines. Be careful, don't cut on any of the dotted lines or you'll mess it up!

3. Score the Model
This is done by dragging a scoring tool — or something not quite sharp enough to cut through the paper, such as a non-serrated butter knife — along the dotted lines. Scoring makes it easier to create precise folds.

4. Glue it All Together
Apply adhesive to the tabs labeled "glue." The thicker the paper, the stronger glue you will need. Now have Makey strike a pose!

Download the PDF and share your results at makezine.com/go/papercraft-makey!

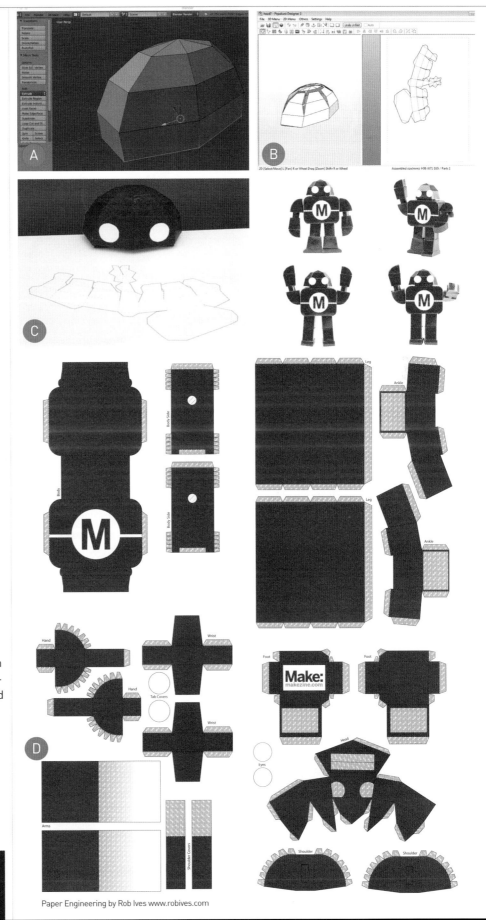

Paper Engineering by Rob Ives www.robives.com

PROJECTS | Fun Zone

Light-Up Copter Launcher

Written by Marc de Vinck ■ Illustrated by Mike Gray

MARC DE VINCK co-developed and co-teaches in the Masters of Engineering in Technical Entrepreneurship (TE) program at Lehigh University in Pennsylvania.

MATERIALS
- » Craft sticks (3)
- » Rubber bands, #33, thin and long (3)
- » Battery, CR2032
- » Standard LED, any color
- » Cereal box or similar cardboard

TOOLS
- » Safety glasses
- » Duct tape
- » Scissors
- » Ruler
- » Needlenose pliers or wire cutters
- » Pencil

An LED illuminates this simple cardboard and craft stick contraption

Did You Know?
Inside the LED, or light-emitting diode, are two types of semiconductor materials — one that is negatively charged and one that is positively charged. When electrical voltage is applied from the battery to these semiconductors, the negative and positive charge carriers are pushed together. When they combine, energy is released in the form of a photon (light particle). This process is called "electroluminescence."

This project is excerpted from *Electricity for Young Makers: Fun and Easy Do-It-Yourself Projects*. Find it at the Maker Shed (makershed.com) and fine bookstores.

THIS TWO-PART PROJECT NOT ONLY LIGHTS UP BUT ALSO FLIES THROUGH THE AIR!

BUILD THE LAUNCHER

1. Gather the materials for this part of the project. You'll need two of the three craft sticks, three #33 rubber bands, and some duct tape. Don't worry too much about the exact size of the rubber bands. Just make sure they are the thin type, stretchy, and fairly long (#33 rubber bands are ⅛" wide and 3½" long).

2. First let's make a chain of 3 rubber bands. Pinch the end of a rubber band to form a little hole about the size of a dime. Next, partially thread another rubber band through the first rubber band, going around it, and back into itself. This makes a knot and a nice connection (Figure Ⓐ). Do the same with the third rubber band, creating a chain that is 3 rubber bands long.

3. Now let's attach the rubber band chain in the same way to a pair of craft sticks. Loop one end of the rubber band chain around a craft stick about ½" from the end. Then feed the other end of the rubber band chain around the wooden craft stick. Do the same with the other end of the rubber band chain to another craft stick (Figure Ⓑ).

4. Next let's make a "V" shape with the craft sticks and secure them at the base with narrow strips of duct tape. Start by tearing a 4"-long piece of duct tape into 3 strips, each about ¼" wide. Wrap one piece of tape around one end of the two sticks to form a "V". The wide part of the "V" should be about 3" apart. Then, wrap a piece of tape vertically between the "V" at the base, followed by wrapping the third piece around the bottom of the "V" again (Figure Ⓒ).

5. And now for the last step. Carefully cut off one side of the middle rubber band in the chain without cutting the knots. We'll use this later to attach our copter. This

GIVE A GIFT.
FULL YEAR ONLY $34.95.
Make:

GIFT FROM

NAME

ADDRESS

CITY   STATE   ZIP

COUNTRY

EMAIL ADDRESS

GIFT TO

NAME

ADDRESS

CITY   STATE   ZIP

COUNTRY

EMAIL ADDRESS *required for access to the digital editions

☐ Please send me my own full year subscription of Make: for $34.95.

We'll send a card announcing your gift. Your recipients can also choose to receive the digital edition at no extra cost.
Price is for U.S. only. For Canada, add $9 per subscription. For orders outside the U.S. and Canada, add $15.

476GS1

NO POSTAGE
NECESSARY
IF MAILED
IN THE
UNITED STATES

BUSINESS REPLY MAIL
FIRST-CLASS MAIL   PERMIT NO. 865   NORTH HOLLYWOOD, CA

POSTAGE WILL BE PAID BY ADDRESSEE

Make:

PO BOX 17046
NORTH HOLLYWOOD CA 91615-9186

will also make the launcher a bit stretchier (Figure D).

That's it! We're done with the launcher.

BUILD THE LED COPTER

1. Gather the electronic components for the build — the CR2032 battery and the LED. Let's take a look at the battery. The positive side is marked with a + sign. See it? On button cell batteries like this one, the negative side isn't usually marked, but the bottom must be negative, right? Batteries have a positive end and a negative end, which allows the flow of electrons to power a circuit.

Now, let's check out the LED. It doesn't have a label for + or –, but it does have two wires, or leads, coming out of it. See how one is longer than the other? The longer wire is the positive (+) lead, and the shorter one is the negative (–) lead (Figure E). Slide the LED onto the battery with the long lead touching the positive (+) side of the battery and the short lead touching the negative (–) side. It should light up.

2. Take the third, and last, craft stick and create a small triangular notch ½" from the end of the stick (Figure F). You can make this notch with needlenose pliers or wire cutters by making small nibbles until it's perfect.

TIP: Be sure it looks just like the one in Figure F. It needs to be shaped like a triangle, and it should have a slight hook shape to it. The notch should cut more than about ⅓ of the way through the craft stick.

3. Cut out two strips of cardboard — the kind cereal boxes are made of — that are ¾" wide by 7" long. Now cut a slight taper into one end of each of these rectangles, about 1" from the end. This makes them a little narrower so you can easily tape them later.

4. Hold both pieces of cardboard together and make a 45° bend about 3" from the tapered end (Figure G).

5. Tape the two strips of cardboard to opposite sides of the craft stick, right below the notch you cut earlier. Make sure the tape overlaps the cardboard and craft stick but doesn't cover the notch. The strips should bend away from the craft stick. These are the "copter blades."

6. There's just one more thing to do — attach the LED and battery to the copter. Slip the LED onto the battery so that the correct leads are touching the + and – sides. The LED should light up. Wrap one more piece of tape around the battery, LED, and craft stick sandwich behind the notch.

TIP: The battery is larger than the craft stick. Make sure it overhangs the craft stick on the side without the notch. If it overhangs by the side with the notch, it will be hard to launch.

We now have a launcher and a glowing LED copter that's ready for action. All we need is a big space, preferably outside, and some darkness! Sure you can try some daytime flights, but where's the fun in that?

Do you want to launch your copter? Of course you do! As all good pilots know, safety comes first, so put on your safety glasses. Now hook the notch at the front of the copter onto the center of the rubber band chain on the launcher. Extend your arms away from your body and face. Aim up toward the sky at about a 45° angle, away from yourself and anyone nearby (Figure H). Pull back on the craft stick of the copter, and let go!

TAKE IT FURTHER

Experiment with different wing shapes and LED configurations. Can you make your copter stay up longer, go farther, or have different colored light patterns? This is also a great project to practice your long-exposure photography. Try taking a picture with a camera mounted on a tripod or stable surface. If you leave the camera shutter open for a long time, you should be able to capture a stream of light like a fireworks show.

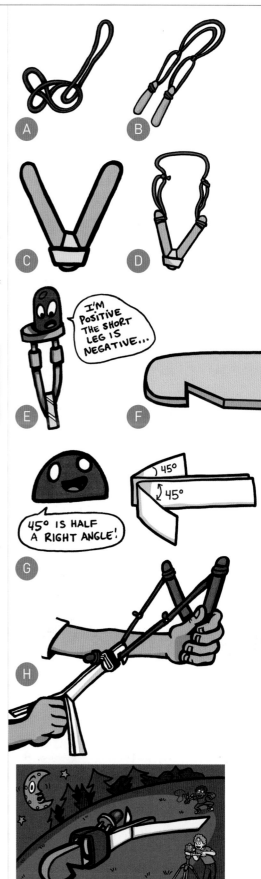

PROJECTS

makezine.com/57

1+2+3 Portable Rally Sign

Written by Paul Spinrad

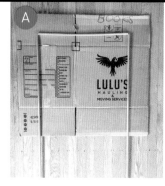

A

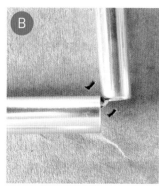

B

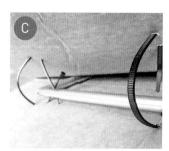

C

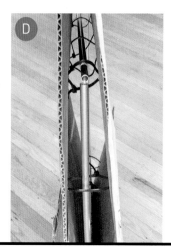

D

WHETHER YOU'RE IN THE DEMONSTRATION OR THE COUNTER-DEMONSTRATION, you need a good sign. Here's how to make a tall and tote-worthy double-sided sign from reusable and reused materials.

1. PUNCH HOLES
With the box flattened, segment the tent pole into a U shape and position it on top of the box, near the top edge of the sign and centered side-to-side (Figure Ⓐ). Using an awl, punch pairs of holes through both layers of the cardboard on either side of the poles (Figure Ⓑ). Turn the box over and widen the holes from the other side.

2. ADD ZIP TIES
Slip the pole between the two sides of the box and string zip ties through the hole pairs, starting from the inside, so the zip tie ends point into the U (Figure Ⓒ). Close the ties around the pole. Reach into the box and pull the ties to tighten them, starting at the top of the sign (Figure Ⓓ).

3. CREATE YOUR SIGN
Decorate paper or cardboard facings, and attach to each side of the sign. That's it!

USE IT
For travel, collapse the tent pole on each side and hold in place with rubber bands. When it's show time, put the tent pole segments together and raise your message high. ⊘

Time Required:
30 Minutes–
1 Hour

Cost:
$0–$35

MATERIALS
» **Shock-corded tent pole**
» **Zip ties**
» **Cardboard box**
» **Poster materials: paper or cardboard, markers or paints**

RAINPROOF VERSION:
» **Coroplast sheets are substituted for the cardboard**
» **Vinyl letter stickers are used instead of markers or paints**

TOOLS
» **Tape, poster putty, or glue**
» **Awl**
» **Rubber bands (2)**

PAUL SPINRAD is a former *Make:* magazine staff editor, writer, maker, and sometime activist. He dresses respectfully and acts serious at political rallies, and it frustrates him when others don't.

We Are All Makers

TOOLBOX | 3D Printer Review

REPLICATOR+
Continuous quality prints indicate the controversial line has fixed its breakdown woes *Written by Matt Stultz*

MACHINE RATING 31

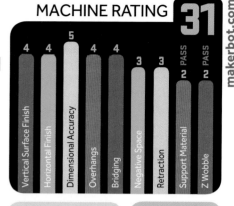

makerbot.com

Category	Score
Vertical Surface Finish	4
Horizontal Finish	4
Dimensional Accuracy	5
Overhangs	4
Bridging	4
Negative Space	3
Retraction	3
Support Material	PASS (2)
Z Wobble	PASS (2)

IN THE EARLY DAYS OF DESKTOP 3D PRINTING, MAKERBOT MADE A NAME FOR ITSELF by offering one of the only kit machines. Their reputation as industry leaders grew until their name became synonymous with 3D printing. Then the trouble came: After losing much of their community goodwill by going closed source, MakerBot released the Gen 5 Replicator. Packed with all kinds of bells and whistles, it had problems, and layoffs ensued as unhappy customers discovered their machines degraded in quality over time. MakerBot is now trying to regain their standing with the release of the Replicator+. While not perfect, the + is a great step up.

REFINED DESIGN
The Replicator+ looks and feels like a 5th Gen at first glance, but a closer inspection reveals that the design has been refined. There is a larger build volume and a much improved bed surface that's not only easily removed, but flexes for smooth print release. The Wi-Fi connectivity actually works, and printing from the mobile app or from the desktop app is simple, although the mobile app could use some improvements. The built-in camera helps you monitor your progress.

While they scored fairly well, the prints still feel slightly off. Layer lines feel chunkier than necessary, seams are not stitched as well as they could be, and top fills are a little sparse even if they score well in our test conditions.

PUT TO THE TEST
For our rating procedure, we run our test prints and score them, but we generally don't reprint them (you can read more about testing and scoring at makezine.com/comparison/3dprinters). However, I've had this machine for months and have reprinted the test models a few times along with other items to see if there would be extruder problems such as those we observed with the 5th Gen. Instead, I actually saw an improvement in my prints, as MakerBot upgraded their software and continued to work on the profiles for their new machine.

PRINTS ARE THE PRODUCT
The Replicator+ seems to indicate that MakerBot is coming back around a bit. The key is that the prints are the really important end product, not the features of the machine. For die-hard MakerBot fans, or those curious about the brand's direction, the Replicator+ is helping right the ship and will make many users happy.

- **MANUFACTURER** MakerBot
- **PRICE AS TESTED** $2,499
- **BUILD VOLUME** 295×195×165mm
- **BED STYLE** Non-heated custom surface
- **FILAMENT SIZE** 1.75mm
- **OPEN FILAMENT?** No (voids warranty)
- **TEMPERATURE CONTROL?** Yes (tool head 235°C max)
- **PRINT UNTETHERED?** Yes (via USB, internal memory, or Wi-Fi)
- **ONBOARD CONTROLS?** Yes (with a scroll wheel and graphic LCD)
- **HOST/SLICER SOFTWARE** MakerBot Print
- **OS** Windows, Mac
- **FIRMWARE** Custom MakerBot
- **OPEN SOFTWARE?** No
- **OPEN HARDWARE?** No
- **MAX DECIBELS** 69.2

PRO TIPS
The convenience of printing remotely from the smartphone app is great, but you still have to physically push the button to kick off your print. You can print from the desktop app while on the same network (or VPN connected anywhere) without the need for a button press.

WHY TO BUY
While not perfect, MakerBot is coming back. If you're hooked on the brand, this is the machine to get.

MATT STULTZ is the 3D printing and digital fabrication lead for *Make:*. He is also the founder and organizer of 3DPPVD and Ocean State Maker Mill, where he spends his time tinkering in Rhode Island.

TEST PRINT

Matt Stultz

BOARD BASICS

Essential **Make:** books for your next board project

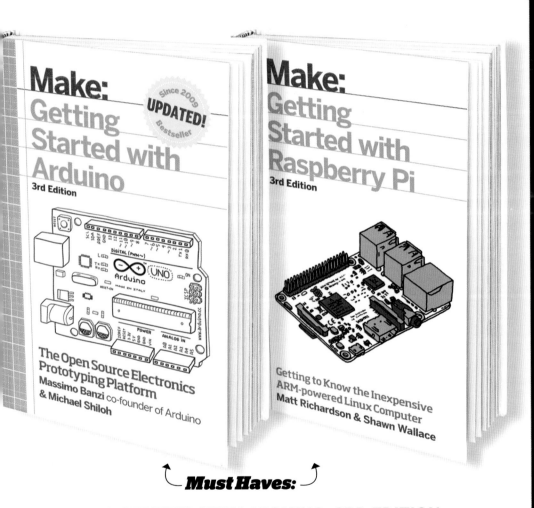

⸺ *Must Haves:* ⸺

GETTING STARTED WITH ARDUINO, 3RD EDITION
The ultimate primer, written by the co-founders of the Arduino Project

GETTING STARTED WITH RASPBERRY PI, 3RD EDITION
The most up-to-date guide available, including info on the Raspberry Pi 3.

MAKE: BOOKS ARE AVAILABLE AT MAKERSHED.COM AND FINE RETAILERS EVERYWHERE

High-Tech Guides:

Make: Action
Learn to monitor and control your world, both online and off.

Make: Arduino Bots and Gadgets
Tackle six embedded open source hardware and software projects!

Make: AVR Programming
Learn how to program AVR microcontrollers directly.

Make: Bluetooth
The ultimate guide to understanding BLE modules, writing code, and wiring circuits.

Make: FPGAs
Design logic that will run on your FPGA and hook up electronic components to create finished projects.

Make: Lego and Arduino Projects
Build projects that extend Mindstorms NXT with open source electronics.

Make: Raspberry Pi and AVR Projects
Four challenging projects perfect for intermediate users.

Make: Sensors
A hands-on primer for using Arduino and Raspberry Pi for real-world monitoring.

"People disagree with me. I just ignore them."
— Linus Torvalds, creator of Linux

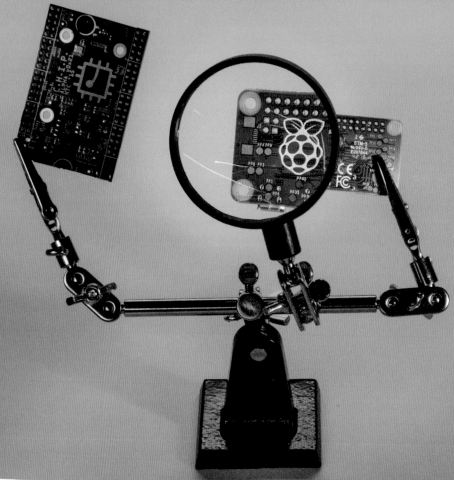

The first Linux book tailored to the needs of the maker movement
Master your Raspberry Pi! Build more projects! Have more fun!

Make: books
Available at **makershed.com** and fine booksellers everywhere

Make: Marketplace

SHOW & TELL

Dazzling projects from inventive makers like you

Sharing what you've made is half the joy of making! Check out these makers on Instagram, and show us your photos by tagging @makemagazine.

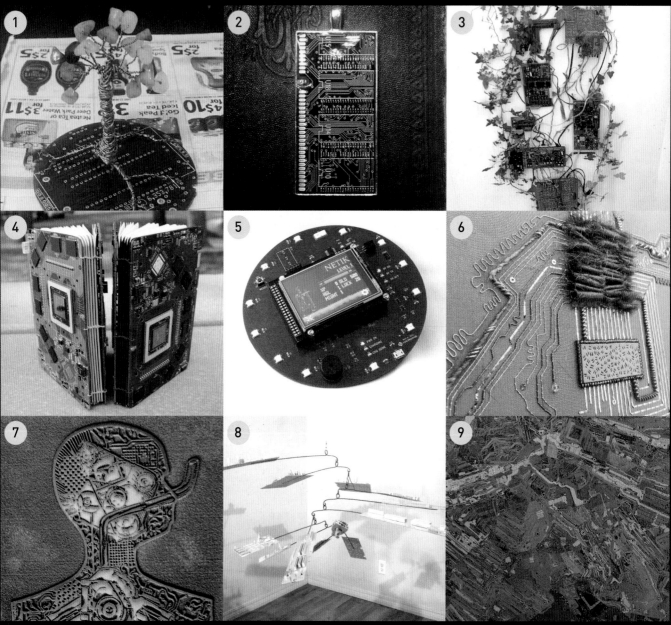

1. **Allie Angelo** (@creative_intoxication) started out making handmade jewelry. The excess and ruined circuit boards she'd take home from her job at an LED lighting company inspired her to build this wire tree with gemstone beads.

2. **Elizabeth Wallace** (@llyzabeth) sets circuit boards into metal frames and cures them with resin to make beautiful pendants.

3. This installation piece by **Harriet Eve Gwyneth Baugh** (@harrietevegwyneth) explores the relationship between nature and modernity. The intertwined circuit boards and natural vines "symbolize the growth of technological progress," she says.

4. **Hollie Olsen** (@itsbitsnbytes) makes both handmade books and upcycled tech jewelry, so these books with circuit board covers were the perfect meeting point between her two hobbies; she hand cuts the circuit boards, and hand stitches the book binding.

5. This hackable electronic badge is the joint effort of a group of makers from San Francisco in celebration of DEFCON's 25th anniversary. **John Adams** (@netik23) shared it, and you can find more info at retina.net/spqr.

6. Textiles student **Livia Papiernik** (@liviapapiernik) used experimental goldwork techniques and unconventional materials to embroider a "circuit board" of thread.

7. **Monty Croasdale**, a student at Astor College for the Arts (@astor_artdept), used a laser cutter to develop this woodcut, inspired by our evolving world in which human jobs are increasingly replaced by automation.

8. This circuit board mobile is the work of artist **Lucas Hicks** (@lhicks00), who wanted to focus on "the transcendence of technology into art," in that "the materials surpass their ordinary limitations of electrical energy and transcend into the realm of kinetic energy."

9. From a distance, this 8-foot-wide abstract piece looks like an aerial view of a rural area, but closer inspection reveals that artist **Wally Dion** (@wally_dion) constructed it from dozens if not hundreds of circuit boards.